Birkhäuser

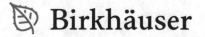

Trends in Mathematics

Trends in Mathematics is a series devoted to the publication of volumes arising from conferences and lecture series focusing on a particular topic from any area of mathematics. Its aim is to make current developments available to the community as rapidly as possible without compromise to quality and to archive these for reference.

Proposals for volumes can be submitted using the Online Book Project Submission Form at our website www.birkhauser-science.com.

Material submitted for publication must be screened and prepared as follows:

All contributions should undergo a reviewing process similar to that carried out by journals and be checked for correct use of language which, as a rule, is English. Articles without proofs, or which do not contain any significantly new results, should be rejected. High quality survey papers, however, are welcome.

We expect the organizers to deliver manuscripts in a form that is essentially ready for direct reproduction. Any version of TEX is acceptable, but the entire collection of files must be in one particular dialect of TEX and unified according to simple instructions available from Birkhäuser.

Furthermore, in order to guarantee the timely appearance of the proceedings it is essential that the final version of the entire material be submitted no later than one year after the conference.

More information about this series at http://www.springer.com/series/4961

Piotr Drygaś • Sergei Rogosin
Editors

Modern Problems in Applied Analysis

Birkhäuser

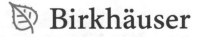

Editors
Piotr Drygaś
Department of Mathematics & Natural Sciences
Rzeszow University
Rzeszów, Poland

Sergei Rogosin
Department of Economics
Belarusian State University
Minsk, Belarus

ISSN 2297-0215 ISSN 2297-024X (electronic)
Trends in Mathematics
ISBN 978-3-030-10242-5 ISBN 978-3-319-72640-3 (eBook)
https://doi.org/10.1007/978-3-319-72640-3

Softcover re-print of the Hardcover 1st edition 2018

Printed on acid-free paper

This book is published under the trade name Birkhäuser, www.birkhauser-science.com by the registered company Springer International Publishing AG part of Springer Nature.
The registered company address is: Gewerbestrasse 11, 6330 Cham, Switzerland

Preface

The book includes extended texts of plenary lectures presented at the third BFA Workshop. The third International Workshop "Boundary Value Problems, Functional Equations and Applications" (BFA3) was held April 20–23, 2016, at the University of Rzeszów, which is the biggest academic institution in Podkarpackie voivodeship situated in the South-East of Poland. The meeting was devoted to recent research in the field of differential and functional equations and complex and real analysis, with a special emphasis on topics related to boundary value problems. The main attention in the reports at BFA3 was paid to different applications of the developed technique. The workshop is organized under the guidance of the International Society for Analysis, its Applications and Computation (ISAAC). The venue of the workshop is related to two previous BFA meetings that took place in Kraków, in 2008 and 2010.

Formally, the content of the book can be divided into two parts. The first part contains the contributions developing the technique related to the scope of the workshop, namely differential and functional equations and real and complex analysis.

The paper by Yu. Alkhutov et al. presents a brief survey of recent new results for the degenerate partial differential equations of p-Laplacian type in a bounded cone. The Dirichlet problem for such equation with the strong nonlinear right part and the Robin problem for such equation with singular nonlinearity in the right part are considered. Such problems are applied to establish mathematical models occurring in reaction-diffusion theory, non-Newtonian fluid theory, non-Newtonian filtration, the turbulent flow of a gas in porous medium, in electromagnetic problems, in heat transfer problems, in Fick's law of diffusion, etc. The aim of the investigation is to study the behavior of weak solutions to the problem in the neighborhood of an angular or conical boundary point of the bounded cone.

The paper by A.O. Celebi deals with a Dirichlet-type problem, known as the Riquier problem, the higher order linear complex differential equation in the unit polydisc of $\mathbb{C}^2$. The constructed Green's function is used to obtain the solution to a model equation with homogeneous boundary conditions.

In the paper by M. Dolfin et al., the models based on the use of the Markov jump processes are discussed. It presents a mathematical description at the microscopic level of the redistribution of individuals in a closed domain featuring, as an example, an elevator. Starting from the microscopic mathematical representation, the corresponding model at the mesoscopic level is also considered. The presented model refers to the position of every separate individual (an agent of the system). Two main classes of probabilities are introduced: one characterizing the influence of the domain, representing the wall and the entrance of the elevator, and another one characterizing the effect of the interactions among agents. The first one is due to two contributions, respectively describing the tendency to concentrate close to the entrance and the tendency to stay close to the boundary. The second one simply represents the tendency to be not too close to other agents.

A method of functional equations for mixed problems for the Laplace equation in multiply connected domains is developed in the paper by V. Mityushev. The considered variant of the method is based on the equivalent reduction of the problem to the Riemann-Hilbert boundary value problem for analytic functions. A problem having applications in composites with a discontinuous coefficient on one of the boundary components is discussed, providing that the domain is a canonical one, namely, lower half-plane with circular holes. A constructive iterative algorithm to obtain an approximate solution is proposed in the form of an expansion in the radius of the holes.

A boundary integral equation method is applied to solve the Riemann-Hilbert problem in the paper by M. Nasser. The solvability of the boundary integral equation with the generalized Neumann kernel is studied for multiply connected domains with smooth boundaries. The boundary integral equation with the generalized Neumann kernel has been used in the study of the Riemann-Hilbert problem in multiply connected domains for special classes of functions. For such special cases, the boundary integral equation with the generalized Neumann kernel has been previously used successfully to compute the conformal mapping onto canonical domains and to solve several boundary value problems such as the Dirichlet problem, the Neumann problem, and the mixed boundary value problem. On the base of the developed method, an alternative proof for the existence and uniqueness of the solution of the general conjugation problem is provided.

The aim of the paper by V. Vasilyev is to describe new interesting examples of nonsmooth manifolds and elliptic pseudo-differential operators acting in functional spaces on such manifolds (e.g., in the Sobolev-Slobodetskii type spaces). Fredholm properties for these operators are studied by factorization methods in the Vishik-Eskin frame, and these are based on the properties of functions of several complex variables.

The paper by M. Zima is devoted to the study of a first-order differential system subject to a nonlocal condition. The goal in this paper is to establish conditions sufficient for the existence of positive solutions when the considered problem is at resonance. The key tool in this approach is Leggett-Williams norm-type theorem for coincidences due to O'Regan and Zima. The paper is concluded with several examples illustrating the main result.

The second part is devoted to applications of solid and fluid mechanics, as well as biomechanics. The central paper of this part by I. Andrianov and V. Mityushev starts with the discussion of the meaning of the term "exact formula" as it is used in the theory of composite materials. The effective properties of composites and review literature on the methods of Rayleigh, Natanzon–Filshtinsky, functional equations, and asymptotic approaches are outlined. In connection with the above methods and new recent publications devoted to composites, the terms *analytical formula, approximate solution, closed form solution, asymptotic formula*, etc. frequently used in applied mathematics and engineering in various contexts are discussed. Though mathematicians give rigorous definitions of exact form solution the term "exact solution" continues to be used too loosely and its attributes are lost. In the present paper, examples of misleading usage of such a term are given. The paper by S. Bock gives an overview of recently developed spatial generalizations of the celebrated Kolosov-Muskhelishvili formulae of elasticity theory using the framework of hypercomplex function theory. Based on these results, a hypercomplex version of the classical Kelvin solution is obtained. For this purpose a new class of monogenic functions with (logarithmic) line singularities is studied, and an associated two-step recurrence formula is proved. Finally, a connection of the function system to the Cauchy-kernel function is established.

A new analytic model describing the behavior of the viscoelastic periodontal ligament after the tooth root translational displacement is suggested in the paper by S. Bosiakov et al. The system of differential equations for the plane-strain state of the viscoelastic periodontal ligament is used as the governing one. The boundary conditions corresponding to the initial small displacement of the root and fixed outer surface of the periodontal ligament in the dental alveolus are employed. A solution is found numerically for the fractional viscoelasticity model assuming that the stresses relaxation in the periodontal ligament after the continuing displacement of the tooth root occurs within about five hours. The obtained results can be used for simulation of the bone remodeling process during orthodontic treatment and for assessment of optimal conditions of the orthodontic load application.

The paper by P. Drygaś is devoted to boundary value problems for elastic problems modeled by the biharmonic equation in two-dimensional composites. It is done in the frame of Mityushev's approach. All the problems are studied via the method of complex potentials. The considered boundary value problems for analytic functions are reduced to functional–differential equations. Applications to calculation of the effective properties tensor are discussed.

The boundary value problem of the theory of elasticity for a finite anisotropic plate with random cracks has been solved in the paper by L. Filshtinskii et al. Stress intensity factors and energy flows near the tips of cracks are determined as a linear functional on solutions to a system of singular integral equations. It is shown that, in case of the normal distribution of the cracks, the statistical characteristics (mathematical expectations and dispersions) of the distraction (stress intensity factors and energy flows) have also the normal law distribution.

In the paper by S. Gluzman and D. Karpeyev, constructive solutions for the effective properties for three problems in the field of random structured media are presented. They are all based on truncated series and on a constructive investigation of their behavior near divergence points where the physical percolation or phase transitions occur. (1) Effective conductivity of 2D conductors with arbitrary contrast parameters is reconstructed from the expansion at small concentrations and of the critical behavior at high concentrations. (2) Effective shear modulus of perfectly rigid spherical inclusions randomly embedded into an incompressible matrix is reconstructed given its expansion at small concentrations and critical behavior. (3) A truncated Fourier expansion to study spontaneous directional ordering in models of planar fully connected suspensions of active polar particles is also employed. The main result is the discovery of a discontinuous, abrupt transition from an ordered to a disordered state. It is a macroscopic effect caused by a mesoscopic self-quenching noise. The relaxation time remains finite at the critical point; therefore the effect of self-quenching is to strongly suppress the critical slowing down and improve the reaction time to external stimuli.

An introductory study of gravity-driven Stokesian flow past the wavy bottom, based on Adler's et al. papers, is proposed by R. Wojnar and W. Bielski. In examples the waviness is described by a sinus function and its amplitude is small, up to $O(\varepsilon^2)$. A correction to Hagen-Poiseuille's type free-flow solution is found. A contribution of capillary surface tension is discussed.

The book is addressed to those who are interested in the development and the real use of the innovative research results, which can help solve scientific and real life problems.

Rzeszów, Poland
Minsk, Belarus
September 2017

Piotr Drygaś
Sergei Rogosin

Contents

Boundary Value Problems for the Singular p- and $p(x)$-Laplacian Equations in a Cone

Yury Alkhutov, Mikhail Borsuk, and Sebastian Jankowski

Abstract In this paper we describe briefly recent new results about the degenerate equations of the p-Laplacian type in a bounded cone. We shall consider the Dirichlet problem for such equation with the strong nonlinear right part as well as the Robin problem for such equation with singular nonlinearity in the right part. Such problems are mathematical models occurring in reaction-diffusion theory, non-Newtonian fluid theory, non-Newtonian filtration, the turbulent flow of a gas in porous medium, in electromagnetic problems, in heat transfer problems, in Fick's law of diffusion et al. The aim of our investigations is the behavior of week solutions to the problem in the neighborhood of an angular or conical boundary point of the bounded cone. We establish sharp estimates of the type $|u(x)| = O(|x|^{\varkappa})$ for the weak solutions u of the problems under consideration.

Keywords Degenerate equations • p-Laplacian type • bvp with strong nonlinear right part • Angular or conical boundary point of the bounded cone • Sharp estimate for week solution

Mathematics Subject Classification (2010) Primary 35J92, 35G30; Secondary 35P20, 76A05

Y. Alkhutov
A. G. and N. G. Stoletov Vladimir State University, Gor'kogo St., Vladimir 600000, Russia

M. Borsuk (✉) • S. Jankowski
Department of Mathematics and Informatics, University of Warmia and Mazury in Olsztyn, 10-957 Olsztyn-Kortowo, Poland
e-mail: borsuk@uwm.edu.pl

P. Drygaś, S. Rogosin (eds.), *Modern Problems in Applied Analysis*, Trends in Mathematics, https://doi.org/10.1007/978-3-319-72640-3_1

1 Boundary Value Problems with Constant Nonlinearity Exponent

1.1 The Dirichlet Problem

Let $\mathcal{C}$ be an open cone in $\mathbb{R}^n$, $n \geq 2$ with vertex in the origin $\mathcal{O}$, B_R be the open ball with radius R centred at $\mathcal{O}$ and $\mathcal{C}_R = \mathcal{C} \cap B_R$. Next let $\Omega = \mathcal{C} \cap S^{n-1}$, where S^{n-1} is the unit sphere centered at $\mathcal{O}$, be the domain obtained by the intersection of the cone $\mathcal{C}$ and the unit sphere. We assume that boundary $\partial\Omega$ of $(n-1)$-dimensional domain Ω, is sufficiently smooth (the smoothness property we shall define more exactly later).

Our interest is the studying of the behavior in a neighborhood of the origin $\mathcal{O}$ of solutions to the Dirichlet problem with boundary condition on the lateral surface of the cone:

$$\begin{cases} Au = b(x, u, \nabla u) & \text{in } \mathcal{C}_{R_0}, \\ u|_{\partial\mathcal{C} \cap B_{R_0}} = 0, & 0 < R_0 \leq 1. \end{cases} \tag{DQL}$$

where

$$Au \equiv \operatorname{div}\left(|\nabla u|^{p-2} a(x) \nabla u\right), \quad p = const > 1 \tag{1.1}$$

under assumptions:

(i) $a(x) = \{a_{ij}(x)\}$ is a measurable symmetric matrix that satisfies for almost all $x \in \mathcal{C}_{R_0}$ the uniform ellipticity condition

$$\nu_1 |\xi|^2 \leq \sum_{i,j=1}^{n} a_{ij}(x) \xi_i \xi_j \leq \nu_2 |\xi|^2 \quad \forall\, \xi \in \mathbb{R}^n,\ 0 < \nu_1 \leq \nu_2, \tag{1.2}$$

(ii) the function $b(x, \eta, \xi)$ is a Caratheodory function, that is $b(x, \eta, \xi)$ for almost all $x \in \mathcal{C}_{R_0}$ is continuous with respect to $(\eta, \xi) \in \mathbb{R} \times \mathbb{R}^n$, measurable with respect to x for all $(\eta, \xi) \in \mathbb{R} \times \mathbb{R}^n$ and satisfies the following inequality:

$$|b(x, u, \nabla u)| \leq \frac{\mu}{1 + |u|} |\nabla u|^p, \quad \mu \in [0, \nu_1). \tag{1.3}$$

We define the functions space $W \equiv W_0^{1,p}(\mathcal{C}_R, \partial\mathcal{C} \cap B_R)$ as the Sobolev space of L^p-integrable in $\mathcal{C}_R$ functions, which have the zero trace on $\partial\mathcal{C} \cap B_R$ and all its weak derivatives of first order exist and are L^p-integrable in $\mathcal{C}_R$ for any $R \in (0, R_0)$.

By a weak solution of problem (DQL) is meant a function $u \in W$ that satisfies the integral identity

$$\int\limits_{\mathcal{C}_{R_0}} \left(|\nabla u|^{p-2}\nabla u \cdot \nabla\varphi + b(x.u, \nabla u)\varphi\right) dx = 0$$

for all bounded test functions $\varphi \in W$ vanishing near the spherical part of the boundary of the domain $\mathcal{C}_{R_0}$.

The main goal of this subsection is the finding of sufficient conditions on the continuity character in $\mathcal{C}_{R_0}$ of $a_{ij}(x)$ and on the boundary $\partial\Omega$ smoothness under the fulfilment which an any solution of the considered problem (DQL) behaves in a neighborhood of the cone $\mathcal{C}$ vertex as $O(|x|^{\lambda})$, where λ is the sharp exponent of the rate of the tending to zero as $|x| \to 0$ for solutions of the problem

$$\begin{cases} \Delta_p v \equiv \operatorname{div}(|\nabla v|^{p-2}\nabla v) = 0, & x \in \mathcal{C}_{R_0}, \\ v(x) = 0, & x \in \partial\mathcal{C} \cap B_{R_0}; \\ 0 < R_0 \leq 1. \end{cases} \qquad (p_L)$$

The behavior of solutions to the p-Laplacian problem (p_L) in a neighborhood of $\mathcal{O}$ was studied by Tolksdorf in [13]. To formulate this result we consider the spherical coordinates (r, ω) and we denote by $|\nabla_\omega u|$ the projection of the gradient ∇u on the tangent plane onto the unit sphere at the point ω:

$$\nabla_\omega \psi = \left\{ \frac{1}{\sqrt{q_1}} \frac{\partial\psi}{\partial\omega_1}, \dots, \frac{1}{\sqrt{q_{n-1}}} \frac{\partial\psi}{\partial\omega_{n-1}} \right\},$$

$$|\nabla_\omega \psi|^2 = \sum_{i=1}^{n-1} \frac{1}{q_i} \left(\frac{\partial\psi}{\partial\omega_i}\right)^2, \quad q_1 = 1,\ q_i = (\sin\omega_1 \cdots \sin\omega_{i-1})^2,\ i \geq 2.$$

If we shall find the (p_L) solution in the form $v(r, \omega) = r^{\lambda}\psi(\omega)$, then we obtain for $(\lambda, \psi(\omega))$ the nonlinear eigenvalue problem

$$\begin{cases} -\operatorname{div}_\omega \left((\lambda^2\psi^2 + |\nabla_\omega\psi|^2)^{(p-2)/2}\nabla_\omega\psi\right) = \\ \lambda\left(\lambda(p-1) + n - p\right)(\lambda^2\psi^2 + |\nabla_\omega\psi|^2)^{(p-2)/2}\psi, \quad \omega \in \Omega, \\ \qquad \psi|_{\partial\Omega} = 0, \end{cases} \qquad (NEVP_D)$$

Tolksdorf investigated in 1983 [13] the behavior near $\mathcal{O}$ of solutions to $(NEVP_D)$ and established that if $\partial\Omega \in C^{\infty}$ then there exist one and only one the least positive eigenvalue

$$\lambda > \max\{0, (p-n)/(p-1)\}, \qquad (1.4)$$

and the corresponding eigenfunction $\psi \in C^{\infty}(\overline{\Omega})$ that solves above nonlinear eigenvalue problem ($NEVP_D$). Moreover,

$$\psi > 0 \quad \text{in } \Omega, \ \psi^2 + |\nabla_\omega \psi|^2 > 0 \quad \text{in } \overline{\Omega}$$

and any two positive eigenfunctions are scalar multiples of each other, if they solve problem for this λ. Next he proved that any solution of (p_L) $u(x) = O(|x|^\lambda)$ and λ is the sharp exponent.

Later on, in 1989, Dobrowolski [8] noted that the condition about the boundary smoothness it is possible weaken: above results are valid if $\partial\Omega \in C^{2+\beta}$ and in this case $\psi \in C^{2+\beta}(\overline{\Omega})$.

In the case of the spherical cone $\mathcal{C}$ nonlinear eigenvalue problem ($NEVP_D$) in Ω was studied for the first time in the Krol and Maz'ya work [11].

This question was studied in more detail in chapter 5 [7]; there different other estimates for the smoothness of the solutions are proved and the extensive bibliography is indicated.

Main result is following statement:

Theorem 1.1 *Let u be a weak solution of the problem (DQL), assumptions* **(i)–(ii)** *are satisfied, $\partial\Omega \in C^{2+\beta}$ and λ be the least positive eigenvalue of problem ($NEVP_D$). Let us assume that $M_{R_0/2} = \sup_{x\in\mathcal{C}_{R_0/2}} |u(x)|$ is known (see below). In addition, suppose that $a_{ij}(x)$ satisfy the Lipschitz condition, i.e.*

$$|a_{ij}(x) - a_{ij}(y)| \le L|x-y|; \ \forall x, y \in \mathcal{C}_{R_0}; \quad a_{ij}(0) = \delta_i^j; \ i, j = 1, \dots, n. \tag{1.5}$$

Then there exist $\varrho_0(n, p, L, \Omega) \le R_0/2$ and constant $C_0 > 0$ depending only on $\mu, \lambda, \nu_1, \varrho_0, M_{R_0/2}$ such that

$$|u(x)| \le C_0|x|^\lambda, \quad x \in \mathcal{C}_{\varrho_0}. \tag{1.6}$$

We give only a sketch of the proof. At first we prove, that under conditions (1.2), (1.3) L_∞-a priori estimate holds, i.e. we derive the value $M_{R_0/2} = \sup_{x\in\mathcal{C}_{R_0/2}} |u(x)|$. Moreover, condition (1.3) is sharp. For this we give two counterexamples. Namely, we show that the condition $0 \le \mu < \nu_1$ in (1.3) is essential and if $\mu > \nu_1$ the solution of the problem (DQL) may be unbounded in the cone vertex.

From the boundedness in $\mathcal{C}_R$ for $R < R_0$ of the problem (DQL) solutions follows their the Höder continuity (see the Ladyzhenskaya–Ural'tseva book [12]: Theorem 1.1 chapter 4, §1).

The proof of the main theorem is based on the barrier method and the using of the maximum principle. For this, at first, we prove

Lemma 1.2 *Let assumptions (1.2), (1.5) are satisfied,* $\partial\Omega \in C^{2+\beta}$ *and* λ *be the least positive eigenvalue of problem (NEVP*$_D$*). Then the inequality*

$$Ah \le -C_0(n,p,L,\Omega)r^{(\lambda-1)p-\lambda(1-\delta)} \text{ a.e. in } \mathcal{C}_\varrho,\ \varrho \le r_0(n,p,L,\Omega) \le R_0$$

holds.

For the proof in detail see [3].

1.2 The Robin Problem

Our interest is the studying of the behavior in a neighborhood of the origin $\mathcal{O}$ of solutions to the Robin problem with singular nonlinearity in the right part and with boundary condition on the lateral surface of the cone:

$$\begin{cases} -\Delta_p u + b(u,\nabla u) = f(x), & x \in \mathcal{C}_{R_0}, \\ |\nabla u|^{p-2}\frac{\partial u}{\partial \vec{n}} + \frac{\gamma}{|x|^{p-1}}u|u|^{p-2} = 0, & x \in \partial\mathcal{C} \cap B_{R_0}, \end{cases} \qquad (RQL)$$

where $0 < R_0 \le 1$ and $p = const > 1$, under assumptions:

- **(i)** $1 < p < n$;
- **(ii)** $|f(x)| \le f_0|x|^\beta$, $f_0 \ge 0$, $\beta > \frac{(p-1)^2}{p-1+\mu}\lambda - p$; $\gamma = const > 0$, $\mu \in [0,1)$ and λ is the least positive eigenvalue of problem ($NEVP_R$) (see below);
- **(iii)** the function $b(u,\xi)$ is differentiable with respect to the u,ξ variables in $\mathfrak{M} = \mathbb{R} \times \mathbb{R}^n$ and satisfy in $\mathfrak{M}$ the following inequalities:

$$|b(u,\xi)| \le \delta|u|^{-1}|\xi|^p + b_0|u|^{p-1}, \quad \delta \in [0,\mu);$$

$$\sqrt{\sum_{i=1}^n \left|\frac{\partial b(u,\xi)}{\partial \xi_i}\right|^2} \le b_1|u|^{-1}|\xi|^{p-1}; \quad \frac{\partial b(u,\xi)}{\partial u} \ge b_2|u|^{-2}|\xi|^p;$$

$$b_0 \ge 0,\ b_1 \ge 0,\ b_2 \ge 0;$$

- **(iv)** the spherical region $\Omega \subset S^{n-1}$ is invariant with respect to rotations in S^{n-2}.

We define the functions space

$$\mathfrak{N}^{1,p}_{-1,\infty}(\mathcal{C}_{R_0}) = \Big\{u \mid u(x) \in L_\infty(\mathcal{C}_{R_0}) \text{ and } \int_{\mathcal{C}_{R_0}} \langle r^{-p}|u|^p + |u|^{-1}|\nabla u|^p\}\rangle\, dx < \infty\Big\}.$$

It is obvious that $\mathfrak{N}^{1,p}_{-1,\infty}(\mathcal{C}_{R_0}) \subset W^{1,p}(\mathcal{C}_{R_0})$.

Remark 1.3 If $p > n$, by the Sobolev imbedding theorem, we have $u \in C^{1-\frac{n}{p}}(\mathcal{C}_{R_0})$. Therefore we investigate only $p \in (1,n)$ *(see assumption* **(i)***).*

Definition 1 The function u is called a weak bounded solution of problem (*RQL*) provided that $u(x) \in \mathfrak{N}^{1,p}_{-1,\infty}(\mathcal{C}_{R_0})$ and satisfies the integral identity

$$Q(u,\eta) :\equiv \int\limits_{\mathcal{C}_{R_0}} \left\langle |\nabla u|^{p-2} u_{x_i}\eta_{x_i} + b\,(u,\nabla u)\,\eta\right\rangle dx + \gamma \int\limits_{\partial\mathcal{C}\cap B_{R_0}} r^{1-p} u|u|^{p-2}\eta dS +$$
$$- \int\limits_{\mathcal{C}\cap S_{R_0}} |\nabla u|^{p-2}\frac{\partial u}{\partial r}\eta d\Omega_d = \int\limits_{\mathcal{C}_{R_0}} f\eta dx. \qquad (II)$$

for all $\eta(x) \in \mathfrak{N}^{1,p}_{-1,\infty}(\mathcal{C}_{R_0})$, where S_{R_0} is the sphere with radius R_0 centred at $\mathcal{O}$. Main result is the following statement:

Theorem 1.4 *Let u be a weak bounded solution of problem* (*RQL*), $M_0 = \sup\limits_{x\in\mathcal{C}_{R_0}} |u(x)|$ *and λ be the least positive eigenvalue of problem (NEVP$_R$) (see below). Suppose assumptions* **(i)–(iv)** *are satisfied. Then there exist $d > 0$ depending only on $p, n, \lambda, (\mu - \delta), b_0$ and constant $C_0 > 0$ depending only on $\lambda, d, M_0, p, n, (\mu - \delta), b_0, f_0$, such that*

$$|u(x)| \le C_0|x|^{\varkappa}, \ \ \varkappa = \frac{p-1}{p-1+\mu}\lambda; \quad \forall x \in \overline{\mathcal{C}_{R_0}}. \qquad (1.7)$$

To prove the main result we shall consider the nonlinear eigenvalue problem for $\psi(\omega) \in C^2(\Omega) \cap C^1(\overline{\Omega})$:

$$\begin{cases} -\mathrm{div}_\omega\left((\lambda^2\psi^2 + |\nabla_\omega\psi|^2)^{(p-2)/2}\nabla_\omega\psi\right) = \\ \lambda\left(\lambda(p-1)+n-p\right)(\lambda^2\psi^2 + |\nabla_\omega\psi|^2)^{(p-2)/2}\psi, & \omega\in\Omega, \\ (\lambda^2\psi^2 + |\nabla_\omega\psi|^2)^{(p-2)/2}\frac{\partial\psi}{\partial\vec{\nu}} + \gamma\left(\frac{p-1+\mu}{p-1}\right)^{p-1}\cdot\psi|\psi|^{p-2} = 0, & \omega\in\partial\Omega, \end{cases} \qquad (NEVP_R)$$

where $\vec{\nu}$ denotes the exterior normal to $\partial\mathcal{C}$ at points of $\partial\Omega$.

We give only the proof sketch. The proof of the main theorem is based on the barrier method and the using of the comparison principle. At first, we derive inequality (1.4) for the least positive eigenvalue of (*NEVP$_R$*).

If we rename $\omega = \omega_1, \omega' = (\omega_2, \cdots, \omega_{n-1})$ then, by assumption **(iv)**, we can consider that $\psi(\omega_1, \omega')$ do not depends on ω'. Therefore our problem (*NEVP$_R$*) is equivalent to the following problem:

$$\begin{cases} \left(\lambda^2\psi^2 + (p-1)\psi'^2\right)\psi''(\omega) + (n-2)\cot\omega\left(\lambda^2\psi^2 + \psi'^2\right)\psi'(\omega) + \\ \lambda\left(\lambda(2p-3)+n-p\right)\psi'^2\psi(\omega) + \lambda^3\left(\lambda(p-1)+n-p\right)\psi^3(\omega) = 0, \\ \omega\in\left(-\frac{\omega_0}{2}, \frac{\omega_0}{2}\right), \\ \pm(\lambda^2\psi^2 + \psi'^2)^{(p-2)/2}\psi'(\omega) + \gamma\left(\frac{p-1+\mu}{p-1}\right)^{p-1}\cdot\psi(\omega)|\psi(\omega)|^{p-2}\Big|_{\omega=\pm\frac{\omega_0}{2}} = 0. \end{cases} \qquad (OEVP)$$

Properties of the (*OEVP*) Eigenfunction $\psi(\omega) \in C^2\left(-\frac{\omega_0}{2}, \frac{\omega_0}{2}\right) \cap C^1\left[-\frac{\omega_0}{2}, \frac{\omega_0}{2}\right]$

First of all we note that any two eigenfunctions are scalar multiples of each other, if they solve problem for the same λ. Therefore without loss of generality we can assume $\psi\left(\frac{\omega_0}{2}\right) = 1$.

Next we observe that it is possible two cases either $\psi(-\omega) = -\psi(\omega)$ or $\psi(-\omega) = \psi(\omega)$. At first, we consider the case $\underline{p = n = 2}$. Then problem (*OEVP*) takes the form

$$\begin{cases} \psi''(\omega) + \lambda^2\psi(\omega) = 0, & \omega \in \left(-\frac{\omega_0}{2}, \frac{\omega_0}{2}\right), \\ \pm\psi'(\omega) + \gamma(1+\mu)\psi(\omega)\Big|_{\omega=\pm\frac{\omega_0}{2}} = 0. & \end{cases}$$

Solving this problem we obtain:

(1) for $\psi(-\omega) = \psi(\omega)$

$$\psi(\omega) = \cos(\lambda\omega), \Rightarrow \psi(0) \neq 0, \ \psi'(0) = 0, \quad \tan\left(\frac{\lambda\omega_0}{2}\right) = \frac{\gamma(1+\mu)}{\lambda};$$

by the graphic method (see Fig. 1), we observe that $0 < \lambda^* < \frac{\pi}{\omega_0}$, where λ^* is the least positive root of the above transcendence equation;

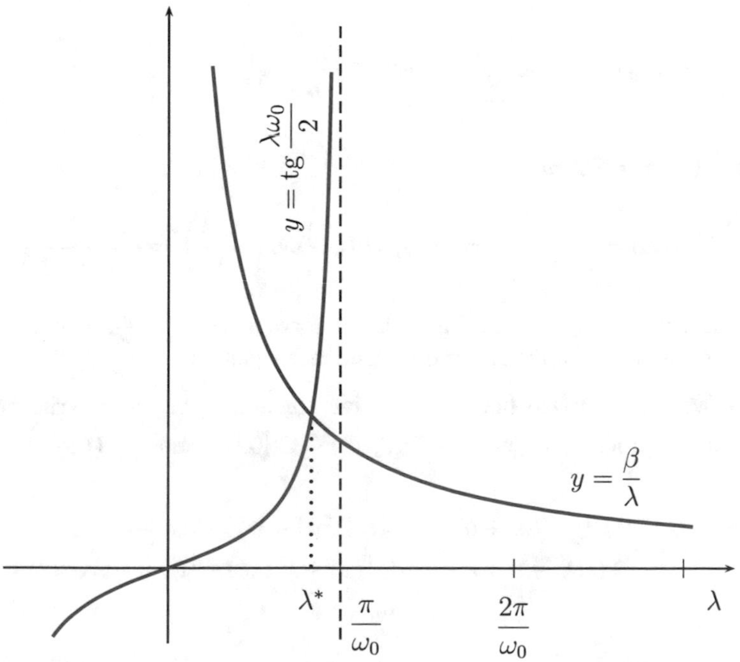

Fig. 1 Solution to transcendence equation $\tan\left(\frac{\lambda\omega_0}{2}\right) = \frac{\gamma(1+\mu)}{\lambda}$

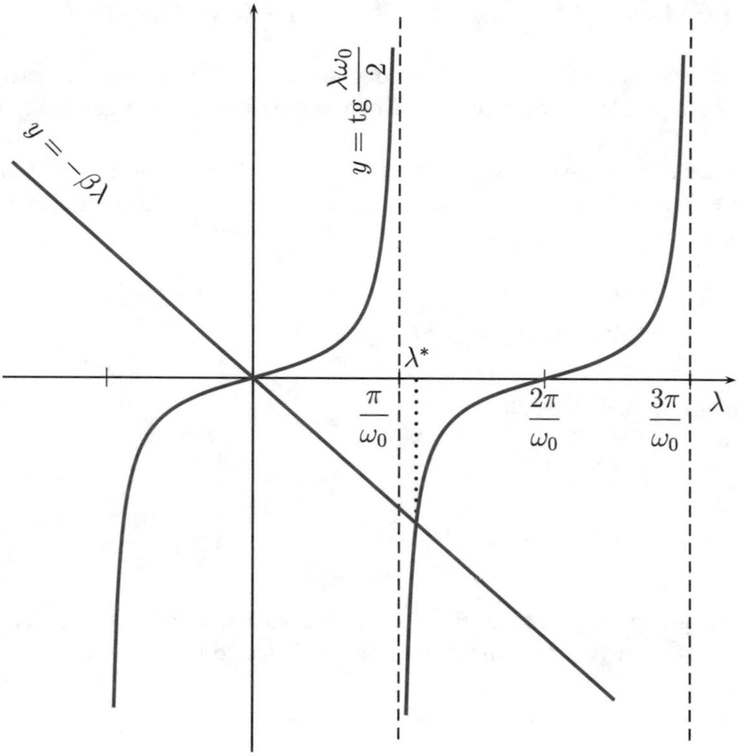

Fig. 2 Solution to transcendence equation $\tan\left(\frac{\lambda\omega_0}{2}\right) = -\frac{\lambda}{\gamma(1+\mu)}$

(2) for $\psi(-\omega) = -\psi(\omega)$

$$\psi(\omega) = \sin(\lambda\omega), \Rightarrow \psi(0) = 0, \quad \tan\left(\frac{\lambda\omega_0}{2}\right) = -\frac{\lambda}{\gamma(1+\mu)};$$

by the graphic method (see Fig. 2), we observe that $\lambda^* > \frac{\pi}{\omega_0}$, where λ^* is the least positive root of the above transcendence equation.

Since we are interested the least positive eigenvalue we can rewrite problem (*OEVP*) as the problem for $\psi(\omega) \in C^2[0, \frac{\omega_0}{2}) \cap C^1[0, \frac{\omega_0}{2}]$ and $\lambda \in \left(0, \frac{\pi}{\omega_0}\right)$:

$$\begin{cases} \left(\lambda^2\psi^2 + (p-1)\psi'^2\right)\psi''(\omega) + (n-2)\cot\omega\left(\lambda^2\psi^2 + \psi'^2\right)\psi'(\omega) + \\ \qquad \lambda\left(\lambda(2p-3) + n - p\right)\psi'^2\psi(\omega) + \lambda^3\left(\lambda(p-1) + n - p\right)\psi^3(\omega) = 0, \\ \qquad\qquad \omega \in \left[0, \frac{\omega_0}{2}\right), \\ \psi'(0) = 0; \quad \psi\left(\frac{\omega_0}{2}\right) = 1; \\ \psi'(\frac{\omega_0}{2})\left\langle\lambda^2 + \psi'^2(\frac{\omega_0}{2})\right\rangle^{\frac{p-2}{2}} = -\gamma\left(\frac{p-1+\mu}{p-1}\right)^{p-1}. \end{cases} \qquad (\psi\lambda)$$

Barrier Function and Eigenvalue Problem (*OEVP*)
We shall study the function $w(r, \omega) \not\equiv 0$ as a solution of the auxiliary problem:

$$\begin{cases} -\Delta_p w = \mu w^{-1}|\nabla w|^p, & x \in \mathcal{C}_{R_0} \\ |\nabla w|^{p-2}\frac{\partial w}{\partial \vec{n}} + \frac{\gamma}{|x|^{p-1}} w|w|^{p-2} = 0, & x \in \partial\mathcal{C} \cap B_{R_0}. \end{cases} \qquad (BFP)$$

By direct calculations, we derive a solution of this problem in the form

$$w = w(r, \omega) = r^{\varkappa}\psi^{\varkappa/\lambda}(\omega), \quad \varkappa = \frac{p-1}{p-1+\mu}\lambda, \qquad (BF)$$

where $(\lambda, \psi(\omega))$ is the solution of the eigenvalue problem (*OEVP*).
Next we prove

Lemma 1.5

$$1 \le \psi(\omega) \le \psi_0 = const(n, p, \lambda, \omega_0), \qquad (1.8)$$

if

- $p = 2,\ n \ge 3$ *and in this case we have* $0 < \lambda < \frac{\pi}{\omega_0}$;
- $2 < p < n$ *and* $\lambda^* < \lambda < \frac{\pi}{\omega_0}$, *where*

$$\lambda^* = \frac{\pi}{\omega_0} \cdot \frac{q\sqrt{\left(\frac{q}{p-1}\right)^2 + \frac{4}{(p-1)(p-2)^2}} - \left(\frac{q^2}{p-1} - \frac{2}{p-2}\right)}{\frac{2}{p-2}\left(\frac{n-p}{p-2} \cdot \frac{\omega_0}{\pi} + 2\right)} > 0, \quad q = \frac{n-p}{p-2} \cdot \frac{\omega_0}{\pi} + 1;$$

- $1 < p < 2 < n$ *and* $0 < \lambda < \frac{\pi}{\omega_0}$.

Lemma 1.6 *The barrier function* $w \in \mathfrak{N}^{1,p}_{-1,\infty}(\mathcal{C}_{R_0})$.
Finally, we show that solution u to (*RQL*) and the barrier function w to (*BF*) satisfy the comparison principle (see §6.2 [5]). By this the proof of main theorem is completed.
For the proof in detail see [6].

2 The Dirichlet Problem with Variable Nonlinearity Exponent

In this section we describe briefly recent new results of our article [2]. We consider the $p(x)$-harmonic equation in $\mathcal{C}_{R_0}$

$$\Delta_{p(x)} u \equiv \mathrm{div}(|\nabla u|^{p(x)-2}\nabla u) = 0$$

with the exponent $p(x)$ that is a measurable function in $\mathcal{C}_{R_0}$, separated from unit and infinity:

$$1 < p_1 \leq p(x) \leq p_2 \quad \forall x \in \mathcal{C}_{R_0}.$$

Our interest is the studying of the behavior in a neighborhood of the origin $\mathcal{O}$ of the $p(x)$-harmonic in $\mathcal{C}_{R_0}$ functions u, satisfying the homogeneous Dirichlet boundary condition on the lateral surface of the cone:

$$\begin{cases} \Delta_{p(x)}u = 0, & x \in \mathcal{C}_{R_0}, \\ u|_{\partial\mathcal{C}\cap B_{R_0}} = 0, & 0 < R_0 \leq 1/2. \end{cases} \tag{$Dp(x)$}$$

We define the functions space

$$W_{loc} = \{w : \; w \in W_0^{1,1}(\mathcal{C}_R, \partial\mathcal{C} \cap B_R), \; |\nabla w|^{p(x)} \in L^1(\mathcal{C}_R)\}$$

for any $R \in (0, R_0)$. Here $W_0^{1,1}(\mathcal{C}_R, \partial\mathcal{C} \cap B_R)$ is the Sobolev space of functions which have the zero trace on $\partial\mathcal{C} \cap B_R$ and all its weak derivatives of first order exist and are L^1-integrable in $\mathcal{C}_R$.

By a solution of our problem is meant a function $u \in W_{loc}$ that satisfies the integral identity

$$\int\limits_{\mathcal{C}_{R_0}} |\nabla u|^{p(x)-2}\nabla u \cdot \nabla\varphi \, dx = 0$$

for all test functions $\varphi \in W_{loc}$ vanishing near the spherical part of the domain $\mathcal{C}_{R_0}$ boundary.

The main goal of this section is the finding of sufficient conditions on the continuity character in $\mathcal{C}_{R_0}$ of exponent $p(x)$ and on the boundary $\partial\Omega$ smoothness under the fulfilment which an any solution of the considered problem behaves in a neighborhood of the cone $\mathcal{C}$ vertex as $O(|x|^\lambda)$, where λ is the sharp exponent of the rate of the tending to zero as $|x| \to 0$ for solutions of the problem

$$\begin{cases} \Delta_{p_0}v == 0, & x \in \mathcal{C}_{R_0}, \\ v|_{\partial\mathcal{C}\cap B_{R_0}} = 0, & 0 < R_0 \leq 1/2 \end{cases} \tag{Dp_0}$$

with $p_0 = p(0)$. If we shall find the (Dp_0) solution in the form $v(r, \omega) = r^\lambda\psi(\omega)$, then we obtain for $(\lambda, \psi(\omega))$ the nonlinear eigenvalue problem $(NEVP_D)$ with $p = p_0$.

The equation with variable exponent $p(x)$ belongs to the wide class of elliptic equations with a nonstandard growth condition and is the Euler equation for variational problems with integrand $|\nabla u|^{p(x)}/p(x)$. Such variational problems were investigated by V. Zhikov in the mid-80s. The important role in the $p(x)$-Laplacian equation theory plays the known as logarithmic condition

$$|p(x) - p(y)| \leq \frac{k_0}{|\ln|x-y||} \quad \forall x, y \in \mathcal{C}_{R_0},\ |x-y| \leq 1/2,$$

that was defined by Zhikov at 1994 [14]. It was introduced initially in order to prove the density of smooth functions in the solutions space. Later, it was found that the logarithmic condition has other numerous consequences. For example, it guarantees the interior Hölder-continuity of solutions, that was established with different methods by Fan (1995) [10] and Alkhutov [1]. The behavior on the boundary of solutions to the Dirichlet problem for $p(x)$-harmonic equation with exponent p, satisfying the logarithmic condition, was studied in the Alkhutov and Krasheninnikova work [4]. In particular, they established that solutions of considered problem are Hölder continuous at the cone $\mathcal{C}$ vertex.

At first, we establish that if $p(x) \in C^0(\overline{\mathcal{C}_{R_0}})$ then $u(x)$ is bounded in $\mathcal{C}_R$ for all $R \in (0, R_0)$; therefore we set

$$M_R = \sup_{\mathcal{C}_R} |u(x)|. \tag{2.1}$$

We obtained the main result under the assumption that exponent p satisfies the Lipschitz condition, i.e.

$$|p(x) - p(y)| \leq L|x-y|, \quad \forall x, y \in \mathcal{C}_{R_0}. \tag{2.2}$$

Theorem 2.1 *Let $\lambda > 0$ be the eigenvalue of above ($NEVP_D$) with $p = p(0) = p_0$. Let the Lipschitz condition (2.2) for $p(x)$ satisfy and $\partial\Omega \in C^{2+\beta}$. Then for any solution of problem ($Dp(x)$) we have*

$$|u(x)| \leq C(\lambda, \beta, p_0, M_{R_0/2})|x|^\lambda, \quad \forall x \in \mathcal{C}_{\rho_0},$$

where $\rho_0 = \rho_0(n, p_0, p_1, L, \Omega, M_{R_0/2}) \leq R_0/2$.

Proof We give only the proof sketch. We apply the barrier method and use the maximum principle. In particular, if $p(x) \equiv const$ then we give the simple proof of the power estimate with the exact exponent for solutions near the cone $\mathcal{C}$ vertex, that early was obtained by Tolksdorf [13], but more complicated way.

We set

$$h = h(w) = \int_0^w \frac{d\tau}{1 + \tau^\delta}, \quad \delta = \min\{1, (2\lambda)^{-1}\},$$

where $w(r, \omega) = r^\lambda \psi(\omega) + r^\gamma = h_1(r, \omega) + h_2(r), \quad \gamma = \lambda(1 + 2\delta)$.

Lemma 2.2 *If $p(x)$ is Lipschitz-continuous (see (2.2)) and $\partial\Omega \in C^{2+\beta}$, then for $\rho \le r_0(n, p_0, p_1, L, \Omega) \le R_0$ we have*

$$\Delta_{p(x)} h \le -C(n, p_0, p_1, L, \Omega) r^{(\lambda-1)p_0-\lambda(1-\delta)}$$

almost everywhere in $\mathcal{C}_\rho$.

We obtain required result from this Lemma, using the maximum principle, because

$$\frac{1}{3} w \le h(w) \le w \quad \text{in } \mathcal{C}_R \; \forall R \in (0, 1).$$

□

The Lipschitz condition play the important role for the construction of the barrier, but it is enough burdensome. Using other methods we can weaken this condition, but only in the case $p_0 = p(0) = 2$, namely if $\partial\Omega \in C^{1+\beta}$ and exponent $p(x) \in C^\alpha(\mathcal{C}_{R_0})$, i.e.

$$|p(x) - p(y)| \le H|x - y|^\alpha, \quad \forall x, y \in \mathcal{C}_{R_0}. \tag{2.3}$$

In this case the sharp estimate near the cone $\mathcal{C}$ vertex for solutions of considered problem is the same as for solutions of similar problem for the Laplace equation. Let μ be the first eigenvalue of the Dirichlet problem for the Laplace-Beltrami operator in Ω. We define

$$\lambda_0 = \frac{2 - n + \sqrt{(n-2)^2 + 4\mu}}{2} \quad \Longrightarrow \quad \lambda_0(\lambda_0 + n - 2) = \mu.$$

Theorem 2.3 *Let (2.3) satisfy, $\partial\Omega \in C^{1+\beta}$ and $p(0) = 2$. Then for any solution $u(x)$ of problem (Dp(x)) we have*

$$|u(x)| \le C(n, p_1, p_2, H, \alpha, \lambda_0, \rho_0, \beta, M_{R_0/2})|x|^{\lambda_0}, \; \forall x \in \mathcal{C}_{\rho_0},$$

where $\rho_0 = \rho_0(n, p_1, p_2, H, \alpha, , \lambda_0, \beta, \Omega, M_{R_0/2}) \le R_0/4$.

Proof The proof is based on integro-differential inequalities (see in detail [2, 5, 7]) and on the pointwise estimate of the gradient modulus. The last estimate follows from the X. Fan results [9]. Namely, he proved that if $p(x) \in C^\alpha$, then $u(x) \in C^{1+\varkappa}(\overline{\mathcal{C}}_{R_0/4} \setminus \{\mathcal{O}\})$. It is used actually the freezing method for $p(x)$.

Putting $U(\rho) = \int\limits_{\mathcal{C}_\rho} |x|^{2-n} |\nabla u|^2 \, dx$, where u is a solution of considered problem, we find that

$$(1 - \rho^{\alpha/4}) U(\rho) \le \frac{\rho}{2\lambda_0} U'(\rho) + \rho^{1+2\lambda_0}, \quad \forall \rho \le r_0 \le R_0/4.$$

where r_0 depends only on n, p, Ω and $M_{R_0/2}$. Hence it follows that

$$U(\rho) \le C\left(2\lambda_0 r_0 + r_0^{-2\lambda_0} U(r_0)\right)\rho^{2\lambda_0}, \quad \forall \rho \in (0, r_0].$$

Further, we derive the local estimate at the boundary for $|u(x)|$.

Proposition 2.4 *Let $\partial\Omega \in C^{1+\beta}$ and (2.3) satisfy. Then for any $\nu > 0$ a solution u of considered problem with $\rho \le R_0/4$ satisfies the inequality*

$$\sup_{x\in\mathcal{C}_{\rho/2,\rho/4}} |u(x)| \le C\Big(\fint_{\mathcal{C}_\rho} |u|^{p_0}\, dx\Big)^{1/p_0} + C\rho^{\nu},$$

where C depends only on n, p, Ω, ν and $M_{R_0/2}$, but $\mathcal{C}_{r_1,r_2} = \mathcal{C} \cap (B_{r_1} \setminus \overline{B}_{r_2})$.
Then stated result follows from this proposition. In fact, because in our case $p_0 = 2$, it is sufficient to use the Wirtinger inequality

$$\int_{\mathcal{C}_\rho} |x|^{-n}|u|^2\, dx \le \mu^{-1} \int_{\mathcal{C}_\rho} |x|^{2-n}|\nabla u|^2\, dx = \mu^{-1}U(\rho),$$

where μ is the first eigenvalue of the Dirichlet problem for the Laplace-Beltrami operator on Ω. The last see in detail [2, 5, 7]. □

References

1. Yu. Alkhutov, The Harnack inequality and the Hölder property of solutions of nonlinear elliptic equations with nonstandard growth condition. Differ. Equ. **33**(12), 1653–1663 (1997)
2. Yu. Alkhutov, M.V. Borsuk, The behavior of solutions to the Dirichlet problem for second order elliptic equations with variable nonlinearity exponent in a neighborhood of a conical boundary point. J. Math. Sci. **210**(4), 341–370 (2015)
3. Yu. Alkhutov, M. Borsuk, The Dirichlet problem in a cone for second order elliptic quasi-linear equation with the p-Laplacian. J. Math. Anal. Appl. **449**, 1351–1367 (2017)
4. Yu. Alkhutov, O. Krasheninnikova, Continuity at boundary points of solutions of quasilinear elliptic equations with a non-standard growth condition. Izv. Math. **68**(6), 1063–1117 (2004)
5. M. Borsuk, *Transmission Problems for Elliptic Second-Order Equations in Non-smooth Domains*. Frontiers in Mathematics (Birkhäuser, Boston, 2010), p. 218
6. M. Borsuk, S. Jankowski, The Robin problem for singular p-Laplacian equation in a cone. Complex Variables Elliptic Equ. (2017). https://doi.org/10.1080/17476933.2017.1307837
7. M. Borsuk, V. Kondratiev, *Elliptic Boundary Value Problems of Second Order in Piecewise Smooth Domains*. North-Holland Mathematical Library, vol. 69 (Elsevier, Amsterdam, 2006), p. 530
8. M. Dobrowolski, On quasilinear elliptic equations in domains with conical boundary points. J. Reine Angew. Math. **394**, 186–195 (1989)
9. X. Fan, Global $C^{1,\alpha}$ regularity for variable exponent elliptic equations in divergence form. J. Differ. Equ. **235**(2), 397–417 (2007)

10. X. Fan, D. Zhao, A class of De Giorgi type and Hölder continuity. Nonlinear Anal. Theory Methods Appl. **36**(3), 295–318 (1999)
11. I.N. Krol', V.G. Maz'ya, The absence of the continuity and Hölder continuity of the solutions of quasilinear elliptic equations near a nonregular boundary. Trudy Moskov. Mat. Obšč (Russian). **26**, 75–94 (1972)
12. O.A. Ladyzhenskaya, N.N. Ural'tseva, *Linear and Quasilinear Elliptic Equations* (Academic Press, New York, 1968)
13. P. Tolksdorf, On the Dirichlet problem for quasilinear equations in domains with conical boundary points. Commun. Partial Differ. Equ. **8**, 773–817 (1983)
14. V.V. Zhikov, On Lavrentiev's phenomenon. Russ. J. Math. Phys. **13**(2), 249–269 (1994)

Exact and "Exact" Formulae in the Theory of Composites

Igor Andrianov and Vladimir Mityushev

Unfairly recognizes advertising, which advertises under the guise of one commodity another product.

—a shortened form of FTC Act, as amended, §15(a), 15 U.S.C.A. §55(a) (Supp. 1938)

The complexity of the model is a measure of misunderstanding the essence of the problem.

—A.Ya. Findlin [20]

Electronic Supplementary Material The online version of this article (https://doi.org/10.1007/978-3-319-72640-3_2) contains supplementary material, which is available to authorized users.

I. Andrianov (✉)
Institut für Allgemeine Mechanik, RWTH Aachen University, Templergraben 64, 52056 Aachen, Germany
e-mail: igor_andrianov@hotmail.com

V. Mityushev
Institute of Computer Sciences, Pedagogical University, ul. Podchorazych 2, Krakow 30-084, Poland
e-mail: mityu@up.krakow.pl

P. Drygaś, S. Rogosin (eds.), *Modern Problems in Applied Analysis*, Trends in Mathematics, https://doi.org/10.1007/978-3-319-72640-3_2

Abstract The effective properties of composites and review literature on the methods of Rayleigh, Natanzon–Filshtinsky, functional equations and asymptotic approaches are outlined. In connection with the above methods and new recent publications devoted to composites, we discuss the terms *analytical formula, approximate solution, closed form solution, asymptotic formula*, etc... frequently used in applied mathematics and engineering in various contexts. Though mathematicians give rigorous definitions of exact form solution the term "exact solution" continues to be used too loosely and its attributes are lost. In the present paper, we give examples of misleading usage of such a term.

Keywords Composite material • Effective property • Closed form solution • Weierstrass function • Eisenstein summation • Asymptotic solution

Mathematics Subject Classification (2010) 74A40, 35C05, 35C20, 33E05, 30E25, 74Q15

1 Introduction

This paper is devoted to the effective properties of composites and review literature concerning analytical formulae. The terms *analytical formula, approximate solution, closed form solution, asymptotic formula* etc... are frequently used in applied mathematics and engineering in various contexts. Though mathematicians give rigorous definitions of exact form solution [21, 31, 51] the term "exact solution" continues to be used too loosely and its attributes are lost. In the present paper, we give examples of misleading usage of such a term. This leads to paradoxical situations when an author *A* approximately solves a problem and an author *B* repeats the result of *A* adding the term "exact solution". Further, the result of *B* is dominated in references due to such an exactness. Examples of such a wrongful usage of the exactness terms are given in the present paper.

We use also the term *constructive solution (method)* understood from the pure utilization point of view. The constructive solution in the present paper means that we have a formula or a precisely described algorithm, perhaps, on the level of symbolic computations. Problems are said to be *tractable* if they can be solved in terms of a closed form expression or of a constructive method. For instance, a typical result from the homogenization theory consists in replacement of a PDE with high oscillated coefficients by an equation with constant coefficients called the effective constants. Next, a boundary value problem is stated to determine the effective constants. Therefore, the homogenization theory rigorously justifies existence of the effective properties. Their determination is a separate question concerning a boundary value problem in a periodicity cell. Though, in some cases, we should not solve a boundary value problem (see for instance the Keller-Dykhne formula [13]).

We think that usage of the terms "constructive solution" and more strong "exact formula" are not acceptable when one writes a formula when its entries could be found from an additional numerical procedure. One may use its own terms but he/she has to be consistent with the commonly used terminology in order to avoid misleading. We want to exclude situations when researchers should necessary add that "our formula is really exact" since it is a spiral way and in the next time they should add that "our formula is really–really exact". Below, we try to discuss the levels of exactness and precise the clear for engineers terminology which should be used in applied mathematics.

The solution of an equation $Ax = b$ can be formally written in the form $x = A^{-1}b$ where A^{-1} is an inverse operator to A.

We say that $x = A^{-1}b$ is a *analytical form solution* if the expression $A^{-1}b$ consists of a finite number of elementary and special functions, arithmetic operations, compositions, integrals, derivatives and series. *Closed form solution* usually excludes usage of series [21, 31, 51].[1]

Asymptotic methods [2, 3] are assigned to analytical methods when solutions are investigated near the critical values of the geometrical and physical parameters. Hence, asymptotic formulae can be considered as analytical approximations.

In order to distinguish our results from others we proceed to classify different types of solutions.

Numerical solution means here the expression $x = A^{-1}b$ which can be treated only numerically. *Integral equation methods* based on the potentials of single and double layers usually give such a solution. An integral in such a method has to be approximated by a cubature formula. This makes it pure numerical, since cubature formulae require numerical data in kernels and fixed domains of integrations.

Series method arises when an unknown element x is expanded into a series $x = \sum_{k=1}^{\infty} c_k x_k$ on the basis $\{x_k\}_{k=1}^{\infty}$ with undetermined constants c_k. Substitution of the series into equation can lead to an infinite system of equations on c_k. In order to get a numerical solution, this system is truncated and a finite system of equations arise, say of order n. Let the solution of the finite system tend to a solution of the infinite system, as $n \to \infty$. Then, the infinite system is called regular and can be solved by the described *truncation method*. This method was justified for some classes of equations in the fundamental book [36]. The series method can be applied to general equation $Ax = b$ in a discrete space in the form of infinite system with infinite number of unknowns. In particular, Fredholm's alternative and the Hilbert-Schmidt theory of compact operators can be applied [36]. So in general, the series method belongs to numerical methods. In the field of composite materials, the series truncation method was systematically used by Guz et al. [32] and many others.

The special structure of the composite systems or application of a low-order truncation can lead to an approximate solution in symbolic form. Par excellence examples of such solutions are due to Rayleigh [62] and to McPhedran et al. [41, 42, 61] where analytical approximate formulae for the effective properties of composites were deduced.

---

[1]However, an integral is accepted. In the same time, Riemann's integral is a limit of Riemann's sums, i.e., it is a series.

Discrete numerical solution refers to applications of the finite elements and difference methods. These methods are powerful and their application is reasonable when the geometry and the physical parameters are fixed. Many experts perceive a pristine computational block (package) as an exact formula: just substitute data and get the result! However, a sackful of numbers is not as useful as an analytical formulae. Pure numerical procedures can fail as a rule for the critical parameters and analytical matching with asymptotic solutions can be useful even for the numerical computations. Moreover, numerical packages sometimes are presented as a remedy from all deceases. It is worth noting again that numerical solutions are useful if we are interested in a fixed geometry and fixed set of parameters for engineering purposes.

Analytical formulae are useful to specialists developing codes for composites, especially for optimal design. We are talking about the creation of highly specialized codes, which enable to solve a narrow class of problems with exceptionally high speed. "It should be emphasized that in problem of design optimization the requirements of accuracy are not very high. A key role plays the ability of the model to predict how the system reacts on the change of the design parameters. This combination of requirements opens a road to renaissance of approximate analytical and semi-analytical models, which in the recent decades have been practically replaced by 'universal' codes" [20].

Asymptotic approaches allow us to define really important parameters of the system. The important parameters in a boundary value problem are those that, when slightly perturbed, yield large variations in the solutions. In other words, asymptotic methods make it possible to evaluate the sensitivity of the system. There is no need to recall that in real problems the parameters of composites are known with a certain (often not very high) degree of accuracy. This causes the popularity of various kinds of assessments in engineering practice. In addition, the fuzzy object oriented (robust in some sense) model [14] is useful to the engineer. Multiparameter models rarely have this quality. "You can make your model more complex and more faithful to reality, or you can make it simpler and easier to handle. Only the most naive scientist believes that the perfect model is the one that represents reality" [30, p. 278]. It should also be remembered that for the construction of multiparameter models it is necessary to have very detailed information on the state of the system. Obtaining such information for an engineer is often very difficult for a number of objective reasons, or it requires a lot of time and money. And the most natural way of constructing sufficiently accurate low-parametric models is to use the asymptotic methods [68].

2 Approximate and Exact Constructive Formulae in the Theory of Composites

In the present section, we discuss constructive methods used in the theory of dispersed composites. The main results are obtained for 2D problems which will be considered in details.

Let ω_1 and ω_2 be the fundamental pair of periods on the complex plane $\mathbb{C}$ such that $\omega_1 > 0$ and $\text{Im}\ \omega_2 > 0$ where Im stands for the imaginary part. The fundamental parallelogram Q is defined by the vertices $\pm\frac{\omega_1}{2}$ and $\pm\frac{\omega_2}{2}$. Without loss of generality the area of Q can be normalized to one. The points $m_1\omega_1 + m_2\omega_2$ $(m_1, m_2 \in \mathbb{Z})$ generate a doubly periodic lattice $\mathcal{Q}$ where $\mathbb{Z}$ stands for the set of integer numbers. Introduce the zeroth cell

$$Q = Q_{(0,0)} = \left\{z = t_1\omega_1 + it_2\omega_2 \in \mathbb{C} : -\frac{1}{2} < t_1, t_2 < \frac{1}{2}\right\}.$$

The lattice $\mathcal{Q}$ consists of the cells $Q_{(m_1,m_2)} = Q_{(0,0)} + m_1\omega_1 + im_2\omega_2$.

Consider N non-overlapping simply connected domains D_k in the cell Q with Lyapunov's boundaries L_k and the multiply connected domain $D = Q \setminus \cup_{k=1}^{N} (D_k \cup L_k)$, the complement of all the closures of D_k to Q (see Figs. 1 and 2). The case of equal disks D_k will be discussed below.

We study conductivity of the doubly periodic composite when the host $D + m_1\omega_1 + m_2\omega_2$ and the inclusions $D_k + m_1\omega_1 + m_2\omega_2$ are occupied by conducting materials. Introduce the local conductivity as the function

$$\sigma(\mathbf{x}) = \begin{cases} 1, & \mathbf{x} \in D, \\ \sigma, & \mathbf{x} \in D_k,\ k = 1, 2, \ldots, N. \end{cases} \tag{2.1}$$

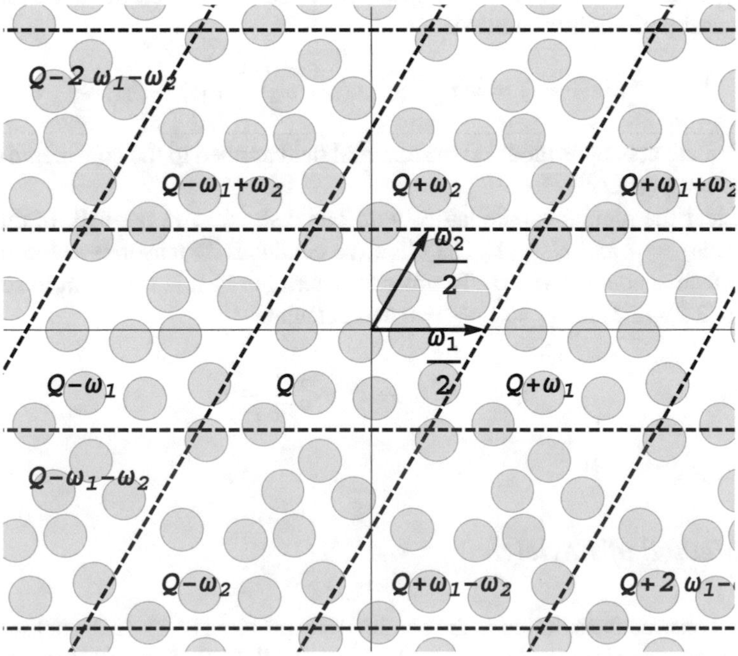

Fig. 1 Doubly periodic composite

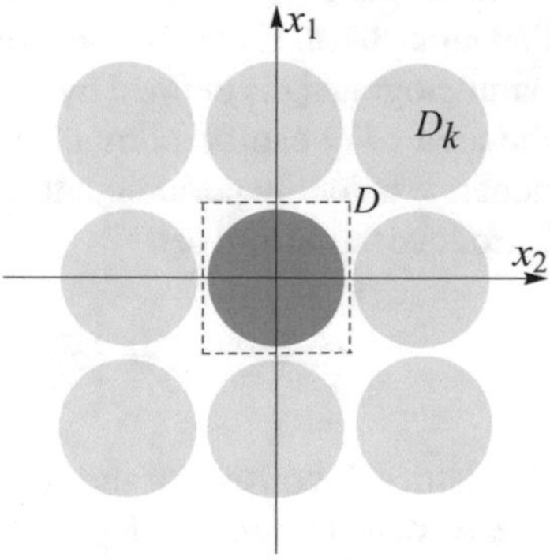

Fig. 2 Square array

Here, $\mathbf{x} = (x_1, x_2)$ is related to the above introduced complex variable z by formula $z = x_1 + ix_2$. The potentials $u(\mathbf{x})$ and $u_k(\mathbf{x})$ are harmonic in D and D_k $(k = 1, 2, \dots, N)$ and continuously differentiable in closures of the considered domains. The conjugation conditions express the perfect contact on the interface

$$u = u_k, \quad \frac{\partial u}{\partial \mathbf{n}} = \sigma \frac{\partial u_k}{\partial \mathbf{n}} \quad \text{on } L_k = \partial D_k, \quad k = 1, 2, \dots, N, \tag{2.2}$$

where $\frac{\partial}{\partial \mathbf{n}}$ denotes the outward normal derivative to L_k. The external field is modelled by the quasi-periodicity conditions

$$u(z + \omega_1) = u(z) + \xi_1, \; u(z + \omega_2) = u(z) + \xi_2, \tag{2.3}$$

where $\xi_{1,2}$ are constants modeled the external field applied to the considered doubly periodic composites.

Consider the regular square lattice with one disk $|z| < r_0$ per cell. In this case, $\omega_1 = 1$, $\omega_2 = i$ and $N = 1$. The effective conductivity tensor is reduced to the scalar effective conductivity σ_e. In order to determine σ_e it is sufficient to solve the problem (2.2)–(2.3) for $\xi_1 = 1$, $\xi_2 = 0$ and calculate

$$\sigma_e = 1 + (\sigma - 1)\, \pi r_0^2 \frac{\partial u_1}{\partial x_1}(0). \tag{2.4}$$

2.1 *Method of Rayleigh*

The first constructive solution to the conductivity problem (2.2)–(2.3) for the regular square array was obtained in 1892 by Lord Rayleigh [62]. The problem is reduced to an infinite system of linear algebraic equations by the series method outlined below. The potentials are expanded into the odd trigonometric series in the local polar coordinates (r, θ)

$$u_1(r, \theta) = C_0 + C_1 r \cos\theta + C_3 r^3 \cos 3\theta + \ldots, \quad r \le r_0, \tag{2.5}$$

$$u(r, \theta) = A_0 + (A_1 r + B_1 r^{-1}) \cos\theta + (A_3 r^3 + B_3 r^{-3}) \cos 3\theta + \ldots. \tag{2.6}$$

The same series can be presented as the real part of analytic functions expanded in the Taylor and Laurent series, respectively,

$$\varphi_1(z) = \varphi_{10} + \varphi_{11} z + \varphi_{13} z^3 + \ldots, \quad r \le r_0, \tag{2.7}$$

$$\varphi(z) = \alpha_0 + \alpha_1 z + \beta_1 z^{-1} + \alpha_3 z^3 + \beta_3 z^{-3} + \ldots, \tag{2.8}$$

where $u_1(z)$ =Re $\varphi_1(z)$ and $u(z)$ =Re $\varphi(z)$. Substitution of the series (2.5)–(2.6) (or the series (2.7)–(2.8)) into (2.2) and selection the terms on $\cos k\theta$ yields an infinite system of linear algebraic equations. Rayleigh treats the quasi-periodicity condition (2.3) as the balance of the multiple sources inside and out of the disk D_1. This leads to the lattice sums $S_m = \sum_{m_1,m_2} (m_1 + im_2)^{-m}$ $(m = 2, 3, \ldots)$, where m_1, m_2 run over all integers except $m_1 = m_2 = 0$. The series S_2 is conditionally convergent, hence, its value depends on the order of summation. Rayleigh uses the Eisenstein summation [70]

$$\sum_{m_1,m_2}^{e} := \lim_{M_2 \to +\infty} \lim_{M_1 \to +\infty} \sum_{m_2=-M_2}^{M_2} \sum_{m_1=-M_1}^{M_1} . \tag{2.9}$$

It is worth noting that Rayleigh (1892) did not refer to Eisenstein (1847) and used Weierstrass's theory of elliptic functions (1856). Perhaps it happened because Eisenstein treated his series formally without study on the uniform convergence introduced by Weierstrass later. Rayleigh used the Eisenstein summation and proved the fascinating formula $S_2 = \pi$ for the square array (see discussion in [71]).

The coefficients of the infinite system are expressed in terms of the lattice sums S_m. This system is written in the next section in the complex form (2.19). Rayleigh truncates the infinite system to get an approximate formula for the effective conductivity (2.4).

Rayleigh extended his method to rectangular arrays of cylinders and to 3D cubic arrays of spherical inclusions. Rayleigh's approach was elaborated by McPhedran with coworkers [41, 42, 61]. For instance, for the hexagonal array Perrins et al. [61] obtained the approximate analytical formula

$$\sigma_e \approx 1 + \frac{2f\varrho}{1 - f\varrho - \frac{0.075422\varrho^2 f^6}{1 - 1.060283\varrho^2 f^{12}} - 0.000076\varrho^2 f^{12}}, \tag{2.10}$$

where $\varrho = \frac{\sigma-1}{\sigma+1}$ denotes the contrast parameter, $f = \pi r_0^2$ the concentration of inclusions. They further developed the method of Rayleigh and applied it to various problems of the theory of composites [11, 53–55, 59].

2.2 Method of Natanzon–Filshtinsky

The next important step in the mathematical treatment of the 2D composites was made by Natanzon [58] in 1935 and Filshtinsky (Fil'shtinskii) in 1964 (see papers [15, 16, 27], his thesis [17], the fundamental books [8, 18, 24–26] and references in [19]). Natanzon and Filshtinsky modified and extended the method of Rayleigh to a 2D elastic doubly periodic problems but without reference to the seminal paper [62].

Rayleigh used the classic Laurent series (2.8) in the domain D of the unit cell and further periodically continued it by the Eisenstein summation (2.9). The main idea of [58] is based on the periodization of the complex potential $\varphi(z)$ without summation over the cells (we take the derivative as in [58])

$$\varphi'(z) = A_0 + A_2\wp(z) + A_4\wp''(z) + A_6\wp^{(iv)}(z) + \dots . \tag{2.11}$$

Filshtinsky used the series

$$\varphi(z) = A_0 z - A_2\zeta(z) - A_4\zeta'(z) - A_6\zeta'''(z) - \dots , \tag{2.12}$$

where $\zeta(z)$ and $\wp(z) = -\zeta'(z)$ are the Weierstrass elliptic functions. Application of series (2.11)–(2.12) allows to avoid the study of the conditionally convergent lattice sum S_2.

Thc Kolosov-Muskhelishvili formulae express the stresses and deformations in terms of two analytic functions. One of the functions has to be doubly periodic, hence, can be presented in the form (2.12). However, a combination of the first and second analytic functions satisfies periodic conditions which yields the following representation for the second function

$$\Phi(z) = B_0 + B_2\wp_1(z) + B_4\wp_1''(z) + \dots , \tag{2.13}$$

where $\wp_1(z)$ is a new function introduced by Natanzon by means of the series

$$\wp_1(z) = -2 \sum_{(m_1,m_2)\in\mathbb{Z}^2\setminus(0,0)} \left[\frac{\overline{P}}{(z-P)^3} - \frac{\overline{P}}{P^3} \right], \tag{2.14}$$

$P = m_1\omega_1 + im_2\omega_2$. Natanzon's function (2.14) was systematically investigated in [26, 58]. In particular, simple expressions of $\wp_1(z)$ and its derivatives in terms of the Weierstrass elliptic functions were established.

Substitution of the series (2.12) into boundary conditions for the conductivity problem with one circular inclusion per cell yields the same Rayleigh's infinite system (see below (2.19)) and low-order in concentration formulae for the effective conductivity. Filshtinsky [15, 27] obtained approximate analytical formulae for the local fields which can yield the effective conductivity after its substitution into (2.4).

In 2012, Godin [22] brought to a close the method of Natanzon-Filshtinsky for conductivity problems having repeated Filshtinsky's analytical approximate formulae for the local fields [15, 27], used (2.4) and arrived at polynomials of order 12 in f for regular arrays. These polynomials are asymptotically equivalent to the approximation (2.10) established in 1978 and can be obtained by truncation from the series, first obtained in 1998, exactly written in the next section. The paper [22] contains the reference to [24] contrary to huge number of papers discussed in Sect. 3.

2.3 *Method of Functional Equations: Exact Solutions*

The methods of Rayleigh and of Natanzon–Filshtinsky are closely related to the method of functional equations proposed in [44] and developed in [31, 45, 51]. Similar functional equations were first applied to boundary value problems by Golusin [23] in 1934.

Roughly speaking, Rayleigh's infinite system is a discrete form of the functional equation. The method of functional equations stands out against other methods in exact solution to 2D problems with one circular inclusion per cell. Moreover, functional equations yield constructive analytical formulae for random composites [31].

In order to demonstrate the connection between the methods of Rayleigh and of functional equations we follow [64] starting from the functional equation

$$\psi_1(z) = \varrho \sum_{m_1,m_2\in\mathbb{Z}}{}' \frac{r^2}{(z-P)^2}\, \overline{\psi_1\left(\frac{r^2}{\overline{z-P}}\right)} + 1, \quad |z| \le r_0, \tag{2.15}$$

where m_1, m_2 run over integers in the sum $\sum'$ with the excluded term $m_1 = m_2 = 0$. Here, $\psi_1(z)$ is the complex flux inside the disk $|z| < r_0$. Let the function $\psi_1(z)$ be expanded in the Taylor series

$$\psi_1(z) = \sum_{m=0}^{\infty} \alpha_m z^m, \quad |z| \le r_0, \tag{2.16}$$

Substituting this expansion into (2.15) we obtain

$$\sum_{m=0}^{\infty} \alpha_m z^m = \varrho \sum_{m=0}^{\infty} r^{2(m+1)} \overline{\alpha_m} \sum_{m_1,m_2 \in \mathbb{Z}}{}' \frac{1}{(z-P)^{m+2}} + 1, \tag{2.17}$$

where the Eisenstein summation (2.9) is used. The function

$$E_l(z) = \sum_{m_1,m_2 \in \mathbb{Z}} \frac{1}{(z-P)^l} \tag{2.18}$$

is called the Eisenstein function of order l [70]. Expanding every function $(z-P)^{-(m+2)}$ in the Taylor series and selecting the coefficients in the same powers of z we arrive at the infinite $\mathbb{R}$-linear algebraic system

$$\alpha_l = \varrho \sum_{m=0}^{\infty} (-1)^m \frac{(l+m+1)!}{l!(m+1)!} S_{l+m+2} r^{2(m+1)} \overline{\alpha_m} + 1, \ l = 0, 1, \ldots. \tag{2.19}$$

This system (2.19) coincides with the system obtained by Filshtinsky. Its real part is Rayleigh's system and Re $\alpha_1 = \frac{\partial u_1}{\partial x_1}(0)$ (see (2.4)).

The functional equation (2.15) is a continuous object more convenient for symbolic computations than the discrete infinite systems (2.19). Instead of expansion in the discrete form described above we solve the functional equation by the method of successive approximations uniformly convergent for any $|\varrho| \leq 1$ and arbitrary r_0 up to touching [31]. Application of (2.4) yields the exact formula for the effective conductivity tensor

$$\sigma_{11} - i\sigma_{12} = 1 + 2\varrho f + 2\varrho^2 f^2 \frac{S_2}{\pi} + \tag{2.20}$$
$$\frac{2\varrho^2 f^2}{\pi} \sum_{k=1}^{\infty} \varrho^k \sum_{m_1=1}^{\infty} \sum_{m_2=1}^{\infty} \cdots \sum_{m_k=1}^{\infty} s_{m_1}^{(1)} s_{m_2}^{(m_1)} \ldots s_{m_k}^{(m_{k-1})} s_1^{(m_k)} \left(\frac{f}{\pi}\right)^{M-k},$$

where $M = 2(m_1 + m_2 + \cdots + m_k)$ and

$$s_k^{(m)} = \frac{(2m+2k-3)!}{(2m-1)!(2k-2)!} S_{2(m+k-1)}. \tag{2.21}$$

The Eisenstein–Rayleigh lattice sums S_m are defined as $S_m = \sum_{P \neq 0} P^{-m}$ and can be determined by computationally effective formulae [31]. The component σ_{22} is calculated by (2.20) where S_2 is replaced by $2\pi - S_2$.

Formula is exact and first was described in 1997 in the papers [48] in the form of expansion on the contrast parameter ϱ. Justification of the uniform convergence for any $|\varrho| \leq 1$ and for an arbitrary radius up to touching disks was established in [47, 48]. The papers [49, 50] contains transformation of the contrast expansion series from [48] to the series (2.20) more convenient in computations.

The relation $S_2 = \pi$ holds for the square and hexagonal arrays (macroscopically isotropic composites). It is worth noting that in this case the terms with $m_1 = m_2 = \ldots = 1$ in the sum over k and the first three terms in (2.20) form a geometric series transforming into the Clausius-Mossotti approximations (Maxwell's formula)

$$\sigma_e \approx \frac{1 + f\varrho}{1 - f\varrho}. \tag{2.22}$$

The series (2.20) can be investigated analytically and numerically. For instance, for the hexagonal array of the perfectly conducting inclusions ($\varrho = 1$) it can be written in the form

$$\begin{aligned} \sigma_e(f) = \quad & 1 + 2f + 2f^2 + 2f^3 + 2f^4 + 2f^5 + 2f^6 \\ & + 2.1508443464271876f^7 + 2.301688692854377f^8 \\ & + 2.452533039281566f^9 + 2.6033773857087543f^{10} \\ & + 2.754221732135944f^{11} + 2.9050660785631326f^{12} \\ & + 3.0674404324522926f^{13} + 3.2411917947659736f^{14} \\ & + 3.426320165504177f^{15} + 3.6228255446669055f^{16} \\ & + 3.8307079322541555f^{17} + 4.049967328265928f^{18} \\ & + 4.441422739726373f^{19} + 4.845994396051242f^{20} \\ & + 5.264540375940583f^{21} + 5.69791875809444f^{22} \\ & + 6.146987621212864f^{23} + 6.6126050439959f^{24} \\ & + 7.135044602470776f^{25} + 7.700073986554016f^{26} \\ & + O(f^{27}). \end{aligned} \tag{2.23}$$

The first 12 coefficients in the expansion of (2.10) in the series in f for $\varrho = 1$ coincide with the coefficients of (2.23).

Application of asymptotic methods to (2.23) yields analytical expressions near the percolation threshold [31]

$$\sigma_e(f) = \alpha(f)\,\frac{F(f)}{G(f)}, \tag{2.24}$$

where

$$\begin{aligned} \alpha(f) = & \frac{4.82231}{\left(\frac{\pi}{\sqrt{12}} - f\right)^{\frac{1}{2}}} - 5.79784 + \\ & 2.13365\left(\frac{\pi}{\sqrt{12}} - f\right)^{\frac{1}{2}} - 0.328432\left(\frac{\pi}{\sqrt{12}} - f\right), \end{aligned} \tag{2.25}$$

$$F(f) = 1.49313 + 1.30576f + 0.383574f^2 + 0.467713f^3 + 0.471121f^4 + 0.510435f^5 + 0.256682f^6 + 0.434917f^7 + 0.813868f^8 + 0.961464f^9 + 0.317194f^{10} + 0.377055f^{11} - 1.2022f^{12} - 0.931575f^{13} \tag{2.26}$$

and

$$G(f) = 1.49313 + 1.30576f + 0.383574f^2 + 0.394949f^3 + 0.4479f^4 + 0.5034f^5 + 0.3033f^6 + 0.2715f^7 + 0.7328f^8 + 0.827239f^9 + 0.25509f^{10} + 0.239752f^{11} - 1.26489f^{12} - f^{13}. \tag{2.27}$$

The function (2.24) is asymptotically equivalent to the polynomial (2.23), as $f \to 0$, and to Keller's type formula [10]

$$\sigma_e(f) = \frac{\sqrt[4]{3}\pi^{\frac{3}{2}}}{\sqrt{2}} \frac{1}{\sqrt{\frac{\pi}{\sqrt{12}} - f}}, \quad \text{as } f \to f_c = \frac{\pi}{\sqrt{12}}. \tag{2.28}$$

The series (2.20) for the hexagonal array when $|\varrho| \leq 1$ yields

$$\sigma_e(f,\varrho) = \frac{1+\varrho f}{1-\varrho f} + 0.150844\varrho^3 f^7 + 0.301688\varrho^4 f^8 + 0.452532\varrho^5 f^9 + 0.603376\varrho^6 f^{10} + 0.75422\varrho^7 f^{11} + \cdots. \tag{2.29}$$

The asymptotic analysis of the local fields when $|\varrho| \to 1$ and $f \to f_c$ was performed in [65] where a criterion of the percolation regime was given, i.e., the domain of validity of (2.29). The following formula was established in [31]

$$\sigma_e(f,\varrho) = \sigma^*(f,\varrho)\,\frac{U(f,\varrho)}{W(f,\varrho)}, \tag{2.30}$$

where

$$\sigma^*(f,\varrho) = \left(1 + \frac{\varrho f}{f_c}\right)^{\frac{1}{2}} \left(1 - \frac{\varrho f}{f_c}\right)^{-\frac{1}{2}}, \tag{2.31}$$

$$U(f,\varrho) = -\varrho^{11} f^{11} - 0.660339\varrho^9 f^9 + \varrho^7(1.44959f^4 - 0.232667)f^7 + \varrho^5(1.59231f^9 + 0.457218f^5) + \varrho^3(1.99365f^7 + 2.10888f^3) + 11.8598\varrho f + 26.4332 \tag{2.32}$$

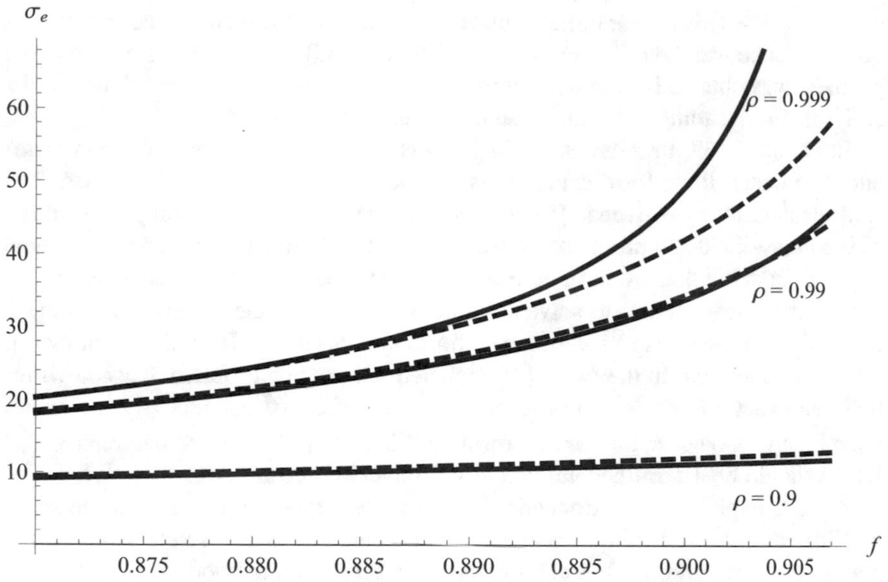

Fig. 3 Effective conductivity for the hexagonal array $\sigma_e(f, \varrho)$ calculated with (2.30)–(2.33) (solid line) and (2.10) (dashed line)

and

$$\begin{aligned} W(f, \varrho) = \varrho^7 f^7 (0.232667 - 1.44959 f^4) + \varrho^{11} f^{11} + 0.660339 \varrho^9 f^9 + \\ \varrho^5(-1.59231 f^9 - 0.457218 f^5) + \varrho^3(-1.99365 f^7 - 2.10888 f^3) - \\ 11.8598 \varrho f + 26.4332. \end{aligned} \tag{2.33}$$

The function (2.30) is asymptotically equivalent to (2.29), as $(\varrho f) \to 0$, and to the percolation regime. Formulae (2.30)–(2.33) and (2.10) for high concentrations are compared in Fig. 3.

3 "Exact Solutions"

Perhaps, the first too magnified usage of the term "exact solution" in the theory of composites began in 1971 with Sendeckyj's paper [67] in the first issue of Journal of Elasticity where "an exact analytic solution is given for a case of antiplane deformation of an elastic solid containing an arbitrary number of circular cylindrical inclusions". Actually, as it is written in this paper, a method of successive approximations based on the method of images was applied to find an approximate solution in analytical form. Sendeckyj's method is a direct application of the generalized alternating method of Schwarz developed by Mikhlin [43, Russian edition in 1949] to circular finitely connected domains. This method does not always converge. The method of Schwarz was modified in [45] where exact[2] solution of the problem was obtained in the form of the Poincaré type series [56]. Its relation to the series method leading to infinite systems is described in [46].

In the book [39, Introduction], Kushch asserts that his solutions are "complete" and "the exact, finite form expressions for the effective properties" obtained. The same declarations are given in [37, 38]. For instance, "exact" and "complete solution of the many-inclusion problem" is declared in [40]. Below [40, Sec. 5] the authors explain that "solution we have derived is asymptotically exact. It means that to get the exact values, one has to solve a whole infinite set of linear equations". In [39, Sec. 9.2.2], difficulties in solution to infinite systems arisen for a finite number of disks are described. In the paper [38] devoted to the same problem, Kushch writes that "an exact and finite form expression of the effective conductivity tensor has been found". However, the "exact" formula (62) from [38] contains parameters $F_n^{(p)}$ which should be determined numerically from an infinite system.

As it is explained in Introduction to the present paper, a regular (in the sense of [36]) infinite system of linear algebraic equations is a discrete form of the continuous Fredholm's integral equation. The truncation method is applied to the both discrete and continuous Fredholm's equations refers to numerical methods in applied mathematics, since a special data set of geometrical and physical parameters is usually taken to get a result. Then, following logical reasoning and Kushch's declarations we should use the term "exact solution" for numerical solutions obtained from Fredholm's equations that essentially distorts the sense of exact solution.

Another question concerns the declaration "complete solution" [37–40]. The completely solved problem should not be investigated anymore. Such a declaration by Kushch ignores exact solutions described in Sect. 2.3 including the exact formulae obtained before, c.f. [45]. It is demonstrated in Fig. 4 that the "complete solution" by Kushch is no longer complete for high concentrations.

Beginning from 2000 Balagurov [4, 5, 7] has been applying the method of Natanzon and Filshtinsky described above in Sect. 2.2 (after exact solution to the problem for regular array of disks in 1997–1998, see Sect. 2.3) without references to them. The main difference between [4, 5, 7] and [58], [18, 24–26] lies in the terminology. Balagurov used the terminology of conductivity governed by Laplace's equation (harmonic functions). Natanzon and Filshtinsky used the terminology of elasticity and heat conduction governed by bi-Laplace's (biharmonic functions) and by Laplace's equation, respectively. Moreover, Filshtinsky separated in his works plane and antiplane elasticity which is equivalent to a separate consideration of bi-Laplace's and Laplace's equation.

For instance, the paper [6] is devoted to application of the Natanzon-Filshtinsky representation (2.12) to the square array of circular cylinders. Exactly having repeated [18, 24–26, 58], Balagurov and Kashin obtained the complex infinite

[2]Not "exact".

system (2.19) written in the paper [6] as (A6). Further, Balagurov and Kashin [6, formulae (21) and (27)] write the effective conductivity up to $O(f^{13})$. This formula from [6] is asymptotically equivalent up to $O(f^{13})$ to formula (14) obtained in the earlier paper [61]. Balagurov's results are not acceptable for high concentrations as displayed in Fig. 4.

Parnell and Abrahams [60] derived "new expressions for the effective elastic constants of the material ... in simple closed forms (3.24)–(3.27)". They "not appealed directly to the theory of Weierstrassian elliptic functions in order to find the effective properties". Actually, they repeated fragments of Eisenstein's approaches (1848) in [60, Sec. 4.1]. Therefore, the same Natanzon-Filshtinsky method was applied to regular arrays of circular cylinders in [60].[3]

A series of papers beginning form 2000, see [29, 63, 66] and references therein, contains a wide set of "exact formulae" for the effective constants hardly accepted as constructive following the lines of Sect. 1. Some of them [12, 28, 63] have the misleading title "Closed-form expressions for the effective coefficients ...". In the paper [29], "the local problems are solved for the case of fiber reinforced composite and the exact-closed formulae for all overall thermoelastic properties" etc. The main results are obtained by the method of Natanzon–Filshtinsky (1935, 1964) without references to it. As in the previous papers, the series (2.11)–(2.12) are substituted into the boundary conditions and Rayleigh's type system of linear algebraic equations is obtained (see for details Supplement to this paper).

Let us consider the recent paper [57] where the 2D square array of circular cylinders and the 3D simple cubic array of spheres of two different radii R_s and R_l were considered when the ratio R_s to R_l was infinitesimally small. The authors declare that "the interaction of periodic multiscale heterogeneity arrangements is exactly accounted for by the reiterated homogenization method. The method relies on an asymptotic expansion solution of the first principles applied to all scales, leading to general rigorous expressions for the effective coefficients of periodic heterogeneous media with multiple spatial scales". This declaration is not true, because a large particle does not interact with another particle of vanishing size. An effective medium approximation valid for dilute composites [52] was actually applied in [57] as the reiterated homogenization theory. As a consequence, formulae (18) and (19) from [57] can be valid to the second order of concentration [52] and additional numerically calculated "terms" are out of the considered precision. The considered problem refers to the general polydispersity problem discussed in 2D statement in [9]. It is not surprisingly that the description of the polydispersity effects in [9] and [57] are different since the "exact" formula from [57] holds only in the dilute regime. This is the reason why the effective conductivity is less than the

[3]The most unexpected paper concerning doubly periodic functions belongs to Wang & Wang [69] where in this one paper (i) rediscovered Weierstrass's functions [69, (2.29) and (2.30)], (ii) rediscovered Eisenstein's summation approach [69, (2.18)], and (iii) solved a boundary value problem easily solved by means of the standard conformal mapping.

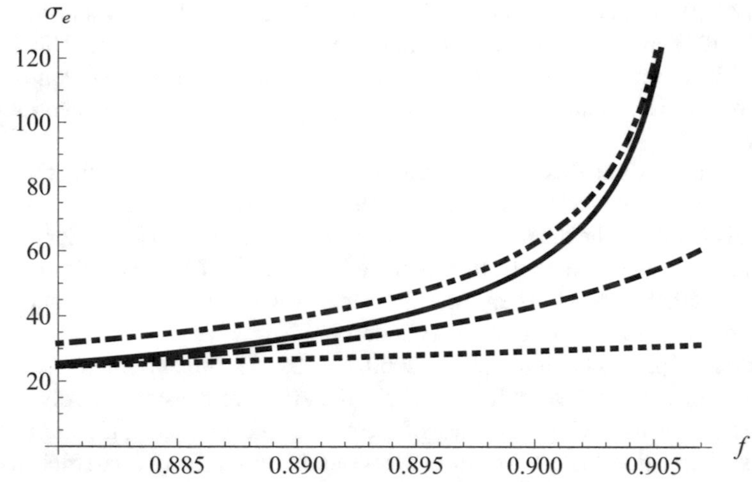

Fig. 4 Effective conductivity for the hexagonal array. Different expressions for σ_e calculated with (2.24)–(2.27) (solid line), (2.10) (dashed line), (2.28) (dot-dashed line). Formula [6, (21) and (27)] is presented by dotted line. Other numerical results from the works discussed in Sect. 3 are presented in Supplement. One can see that they are located on the dotted line or below it

conductivity of matrix reinforced by higher conducting inclusions in Figs. 3 and 7 of [57] for high concentrations.

4 Conclusion

To ensure that our statements were not unfounded, in Supplement we subjected one of the results of [28] to a detailed analysis. We can draw the following conclusions.

1. Authors' claims on closed-form expressions for the effective coefficients, true for any parameter values, are absolutely unjustified.
2. The results obtained by them in this particular case are correct, but they have a very limited field of applicability and are presented in a form that does not allow direct use. Only after making considerable efforts and becoming, in fact, the co-author of the paper, the reader is able to get simple formulae. As a result, reader is convinced that this is a minor modification (and not in the direction of improvement) of well-known formulae.

The question arises of the usefulness of such works—they are clearly inaccessible to the engineer, they do not contain new mathematical results, the field of applicability has not been evaluated in any way.

The general conclusion is that the expressions "exact" or "closed-form" solutions are very obligatory. The authors of the articles should not use them in vain, and reviewers and editors must stop unreasonable claims.

Note also numerous attempts to reinvent the wheel, which look especially strange in time of Google's Empire and scientific social networks (Research Gate, etc.)

In fact, an infinite system of linear algebraic equations contains a remarkable amount of information, which the investigator should do his best to extract (see, c.f., [36]). However, we can not usually see this in papers with "exact solutions".

We note one more aspect of the application of analytic, in particular, asymptotic, methods. The significant disadvantage inherent in them—the local nature of the results obtained—is overcome by modern methods of summation and interpolation (Padé approximants, two-point Padé approximants, asymptotically equivalent functions, etc. [1, 31, 33–35]). Example from our paper: formulae (2.24)–(2.27) are deduced by asymptotic matching (2.23) near $f = 0$ and (2.28) near $f = f_c$.

Acknowledgements Authors thanks Dr Galina Starushenko for fruitful discussions and providing the working notes with the calculation data and figures presented in Supplement.

References

1. I.V. Andrianov, J. Awrejcewicz, New trends in asymptotic approaches: summation and interpolation methods. Appl. Mech. Rev. **54**, 69–92 (2001)
2. I.V. Andrianov, H. Topol, Asymptotic analysis and synthesis in mechanics of solids and nonlinear dynamics (2011). arxiv.org/abs/1106.1783.
3. I.V. Andrianov, J. Awrejcewicz, B. Markert, G.A. Starushenko, Analytical homogenization for dynamic analysis of composite membranes with circular inclusions in hexagonal lattice structures. Int. J. Struct. Stab. Dyn. **17**, 1740015 (2017)
4. B.Ya. Balagurov, Effective electrical characteristics of a two-dimensional three-component doubly-periodic system with circular inclusions. J. Exp. Theor. Phys. **92**, 123–134 (2001)
5. B.Ya. Balagurov, *Electrophysical Properties of Composite: Macroscopic Theory* (URSS, Moscow, 2015) (in Russian)
6. B.Ya. Balagurov, V.A. Kashin, Conductivity of a two-dimensional system with a periodic distribution of circular inclusions. J. Exp. Theor. Phys. **90**, 850–860 (2000)
7. B.Ya. Balagurov, V.A. Kashin, Analytic properties of the effective dielectric constant of a two-dimensional Rayleigh model. J. Exp. Theor. Phys. **100**, 731–741 (2005)
8. D.I. Bardzokas, M.L. Filshtinsky, L.A. Filshtinsky, *Mathematical Methods in Electro-Magneto-Elasticity* (Springer, Berlin, 2007)
9. L. Berlyand, V. Mityushev, Increase and decrease of the effective conductivity of a two phase composites due to polydispersity. J. Stat. Phys. **118**, 481–509 (2005)
10. L. Berlyand, A. Novikov, Error of the network approximation for densely packed composites with irregular geometry. SIAM J. Math. Anal. **34**, 385–408 (2002)
11. L.C. Botten, N.A. Nicorovici, R.C. McPhedran, C.M. de Sterke, A.A. Asatryan, Photonic band structure calculations using scattering matrices. Phys. Rev. E **64**, 046603 (1971)
12. J. Bravo-Castillero, R. Guinovart-Diaz, F.J. Sabina, R. Rodriguez-Ramos, Closed-form expressions for the effective coefficients of a fiber-reinforced composite with transversely isotropic constituents - II. Piezoelectric and square symmetry. Mech. Mater. **33**, 237–248 (2001)
13. A.M. Dykhne, Conductivity of a two-dimensional two-phase system. Sov. Phys. JETP **32**, 63–65 (1971)

14. E.S. Ferguson, How engineers lose touch. Invent. Technol. **8**, 16–24 (1993)
15. L.A. Fil'shtinskii, Heat-conduction and thermoelasticity problems for a plane weakened by a doubly periodic system of identical circular holes. Tepl. Napryazh. Elem. Konstr. **4**, 103–112 (1964) (in Russian)
16. L.A. Fil'shtinskii, Stresses and displacements in an elastic sheet weakened by a doubly periodic set of equal circular holes. J. Appl. Math. Mech. **28**, 530–543 (1964)
17. L.A. Fil'shtinskii, Toward a solution of two-dimensional doubly periodic problems of the theory of elasticity. PhD thesis (Novosibirsk University, Novosibirsk, 1964) (in Russian)
18. L.A. Fil'shtinskii, Physical fields modelling in piece-wise homogeneous deformable solids. Publ. SSU, Sumy (2001) (in Russian)
19. L. Filshtinsky, V. Mityushev, Mathematical models of elastic and piezoelectric fields in two-dimensional composites, in *Mathematics Without Boundaries*, ed. by P.M. Pardalos, Th.M. Rassias. Surveys in Interdisciplinary Research (Springer, New York, 2014), pp. 217–262
20. A.Ya. Findlin, Peculiarities of the use of computational methods in applied mathematics (on global computerization and common sense). Applied Mathematics: Subject, Logic, Peculiarities of Approaches. With Examples from Mechanics, ed. by I.I. Blekhman, A.D. Myshkis, Ya.G. Panovko (URSS, Moscow, 2007), pp. 350–358 (in Russian)
21. F.D. Gakhov, *Boundary Value Problems*, 3rd edn. (Nauka, Moscow, 1977) (in Russian); Engl. transl. of 1st edn. (Pergamon Press, Oxford, 1966)
22. Y.A. Godin, The effective conductivity of a periodic lattice of circular inclusions. J. Math. Phys. **53**, 063703 (2012)
23. G.M. Golusin, Solution to basic plane problems of mathematical physics for the case of Laplace's equation and multiply connected domains bounded by circles (method of functional equations). Matematicheskij zbornik **41**, 246–276 (1934) (in Russian)
24. E.I. Grigolyuk, L.A. Filshtinsky, *Perforated Plates and Shells* (Nauka, Moscow, 1970) (in Russian)
25. E.I. Grigolyuk, L.A. Filshtinsky, *Periodical Piece-Homogeneous Elastic Structures* (Nauka, Moscow, 1991) (in Russian)
26. E.I. Grigolyuk, L.A. Filshtinsky, *Regular Piece-Homogeneous Structures with Defects* (Fiziko-Matematicheskaja Literatura, Moscow, 1994) (in Russian)
27. E.I. Grigolyuk, L.M. Kurshin, L.A. Fil'shtinskii, On a method to solve doubly periodic elastic problems. Prikladnaya Mekhanika (Appl. Mech.) **1**, 22–31 (1965) (in Russian)
28. R. Guinovart-Diaz, J. Bravo-Castillero, R. Rodriguez-Ramos, F.J. Sabina, Closed-form expressions for the effective coefficients of a fibre-reinforced composite with transversely isotropic constituents – I. Elastic and hexagonal symmetry. J. Mech. Phys. Solids **49**, 1445–1462 (2001)
29. R. Guinovart-Diaz, R. Rodriguez-Ramos, J. Bravo-Castillero, F.J. Sabina, G.A. Maugin, Closed-form thermoelastic moduli of a periodic three-phase fiber-reinforced composite. J. Therm. Stresses **28**, 1067–1093 (2005)
30. J. Gleik, *Chaos: Making a New Science* (Viking Penguin, New York, 1987)
31. S. Gluzman, V. Mityushev, W. Nawalaniec, *Computational Analysis of Structured Media* (Elsevier, Amsterdam, 2017)
32. A.N. Guz, V.D. Kubenko, M.A. Cherevko, *Diffraction of Elastic Waves* (Naukova Dumka, Kiev, 1978) (in Russian).
33. A.L. Kalamkarov, I.V. Andrianov, G.A. Starushenko, Three-phase model for a composite material with cylindrical circular inclusions. Part I: application of the boundary shape perturbation method. Int. J. Eng. Sci. **78**, 154–177 (2014)
34. A.L. Kalamkarov, I.V. Andrianov, G.A. Starushenko, Three-phase model for a composite material with cylindrical circular inclusions. Part II: application of Padé approximants. Int. J. Eng. Sci. **78**, 178–219 (2014)
35. A.L. Kalamkarov, I.V. Andrianov, P.M.C.L. Pacheco, M.A. Savi, G.A. Starushenko, Asymptotic analysis of fiber-reinforced composites of hexagonal structure. J. Multiscale Model. **7**, 1650006 (2016)
36. L.V. Kantorovich, V.I. Krylov, *Approximate Methods of Higher Analysis* (Noordhoff, Groningen, 1958)

37. I. Kushch, Stress concentration and effective stiffness of aligned fiber reinforced composite with anisotropic constituents. Int. J. Solids Struct. **45**, 5103–5117 (2008)
38. I. Kushch, Transverse conductivity of unidirectional fibrous composite with interface arc cracks. Int. J. Eng. Sci. **48**, 343–356 (2010)
39. I. Kushch, *Micromechanics of Composites* (Elsevier/Butterworth-Heinemann, Amsterdam/Oxford, 2013)
40. I. Kushch, S.V. Shmegera, V.A. Buryachenko, Elastic equilibrium of a half plane containing a finite array of elliptic inclusions. Int. J. Solids Struct. **43**, 3459–3483 (2006)
41. R.C. McPhedran, D.R. McKenzie, The conductivity of lattices of spheres. I. The simple cubic lattice. Proc. Roy. Soc. London **A359**, 45–63 (1978)
42. R.C. McPhedran, L. Poladian, G.W. Milton, Asymptotic studies of closely spaced, highly conducting cylinders. Proc. Roy. Soc. London **A415**, 185–196 (1988)
43. S.G. Mikhlin, *Integral Equations and their Applications to Certain Problems in Mechanics, Mathematical Physics, and Technology*, 2nd edn. (Pergamon Press, Oxford, 1964)
44. V. Mityushev, Boundary value problems and functional equations with shifts in domains, PhD Thesis, Minsk (1984) (in Russian)
45. V. Mityushev, Plane problem for the steady heat conduction of material with circular inclusions. Arch. Mech. **45**, 211–215 (1993)
46. V. Mityushev, Generalized method of Schwarz and addition theorems in mechanics of materials containing cavities. Arch. Mech. **45**, 1169–1181 (1995)
47. V. Mityushev, Functional equations and its applications in mechanics of composites. Demonstratio Math. **30**, 64–70 (1997)
48. V. Mityushev, Transport properties of regular array of cylinders. ZAMM **77**, 115–120 (1997)
49. V. Mityushev, Steady heat conduction of the material with an array of cylindrical holes in the non-linear case. IMA J. Appl. Math. **61**, 91–102 (1998)
50. V. Mityushev, Exact solution of the $\mathbb{R}$-linear problem for a disk in a class of doubly periodic functions. J. Appl. Funct. Anal. **2**, 115–127 (2007)
51. V.V. Mityushev, S.V. Rogosin, *Constructive Methods for Linear and Nonlinear Boundary Value Problems for Analytic Functions. Theory and Applications*. Monographs and Surveys in Pure and Applied Mathematics (Chapman & Hall/CRC, Boca Raton, 2000)
52. V. Mityushev, N. Rylko, Maxwell's approach to effective conductivity and its limitations. Q. J. Mech. Appl. Math. **66**, 241–251 (2013)
53. A.B. Movchan, S. Guenneau, Split-ring resonators and localized modes. Phys. Rev. B **70**, 125116 (2004)
54. A.B. Movchan, N.V. Movchan, Ch.G. Poulton, *Asymptotic Models of Fields in Dilute and Densely Packed Composites* (World Scientific, London, 2002)
55. A.B. Movchan, N.V. Movchan, R.C. McPhedran, Bloch–Floquet bending waves in perforated thin plates. Proc. Roy. Soc. London **A463**, 2505–2518 (2007)
56. V. Mityushev, Poincaré α-series for classical Schottky groups and its applications, in *Analytic Number Theory, Approximation Theory, and Special Functions*, ed. by G.V. Milovanović, M.Th. Rassias (Springer, Berlin, 2014), pp. 827–852
57. E.S. Nascimento, M.E. Cruz, J. Bravo-Castillero, Calculation of the effective thermal conductivity of multiscale ordered arrays based on reiterated homogenization theory and analytical formulae. Int. J. Eng. Sci. **119**, 205–216 (2017)
58. V.Ya. Natanzon, On stresses in a tensioned plate with holes located in the chess order. Matematicheskii sbornik **42**, 617–636 (1935) (in Russian)
59. N.A. Nicorovici, G.W. Milton, R.C. McPhedran, L.C. Botten, Quasistatic cloaking of two-dimensional polarizable discrete systems by anomalous resonance. Opt. Express **15**, 6314–6323 (2007)
60. W. Parnell, I.D. Abrahams, Dynamic homogenization in periodic fibre reinforced media. Quasi-static limit for SH waves. Wave Motion **43**, 474–498 (2006)
61. W.T. Perrins, D.R. McKenzie, R.C. McPhedran, Transport properties of regular arrays of cylinders. Proc. Roy. Soc. London **A369**, 207–225 (1979)

62. Rayleigh, On the influence of obstacles arranged in rectangular order upon the properties of medium. Phil. Mag. **34**, 481–502 (1892)
63. R. Rodriguez-Ramos, F.J. Sabina, R. Guinovart-Diaz, J. Bravo-Castillero, Closed-form expressions for the effective coefficients of a fibre-reinforced composite with transversely isotropic constituents – I. Elastic and square symmetry. Mech. Mater. **33**, 223–235 (2001)
64. N. Rylko, Transport properties of a rectangular array of highly conducting cylinders. Proc Roy. Soc. London **A472**, 1–12 (2000)
65. N. Rylko, Structure of the scalar field around unidirectional circular cylinders. J. Eng. Math. **464**, 391–407 (2008)
66. F.J. Sabina, J. Bravo-Castillero, R. Guinovart-Diaz, R. Rodriguez-Ramos, O.C. Valdiviezo-Mijangos, Overall behaviour of two-dimensional periodic composites. Int. J. Solids Struct. **39**, 483–497 (2002)
67. G.P. Sendeckyj, Multiple circular inclusion problems in longitudinal shear deformation. J. Elast. **1** 83–86 (1971)
68. A.B. Tayler, *Mathematical Models in Applied Mechanics* (Clarendon Press, Oxford, 2001)
69. Y. Wang, Y. Wang, Schwarz-type problem of nonhomogeneous Cauchy-Riemann equation on a triangle. J. Math. Anal. Appl. **377**, 557–570 (2011)
70. A. Weil, *Elliptic Functions According to Eisenstein and Kronecker* (Springer, Berlin, 1976)
71. S. Yakubovich, P. Drygas, V. Mityushev, Closed-form evaluation of 2D static lattice sums. Proc. Roy. Soc. London **A472**, 20160510 (2016)

On a Hypercomplex Version of the Kelvin Solution in Linear Elasticity

Sebastian Bock

Abstract The article gives an overview about recently developed spatial generalizations of the Kolosov-Muskhelishvili formulae using the framework of hypercomplex function theory. Based on these results, a hypercomplex version of the classical Kelvin solution is obtained. For this purpose a new class of monogenic functions with (logarithmic) line singularities is studied and an associated two step recurrence formula is proved. Finally, a connection of the function system to the Cauchy-kernel function is established.

Keywords Recurrence formulae • Generalized Kolosov-Muskhelishvili formulae • Kelvin solution

Mathematics Subject Classification (2010) Primary 30G35; Secondary 74B05

1 Introduction

The *method of complex stress functions*, which is mainly associated with the work of Kolosov [13] and Muskhelishvili [17], is one of the most fruitful and elegant techniques for solving boundary value problems in the linear theory of elasticity. Here the stresses and deformations of a linear elastic, isotropic, homogeneous body in $\mathbb{C} \cong \mathbb{R}^2$ are represented by the so-called Kolosov-Muskhelishvili formulae

$$\begin{aligned}\sigma_{xx} + \sigma_{yy} &= 2\Big[\,\boldsymbol{\Phi}'(z) + \overline{\boldsymbol{\Phi}'(z)}\,\Big] \;=\; 4\,\mathbf{Re}\big(\boldsymbol{\Phi}'(z)\big),\\ \sigma_{yy} - \sigma_{xx} + 2\boldsymbol{i}\,\tau_{xy} &= 2\Big[\,\bar{z}\,\boldsymbol{\Phi}''(z) + \boldsymbol{\Psi}'(z)\,\Big],\\ 2\mu\,(\mathrm{u} + \boldsymbol{i}\,\mathrm{v}) &= \kappa\,\boldsymbol{\Phi}(z) - z\,\overline{\boldsymbol{\Phi}'(z)} - \overline{\boldsymbol{\Psi}(z)}\end{aligned} \tag{1.1}$$

S. Bock (✉)
Institute of Mathematics/Physics, Bauhaus-Universität Weimar, Coudraystr. 13B, 99423 Weimar, Germany
e-mail: sebastian.bock@uni-weimar.de

P. Drygaś, S. Rogosin (eds.), *Modern Problems in Applied Analysis*, Trends in Mathematics, https://doi.org/10.1007/978-3-319-72640-3_3

in terms of two holomorphic functions $\boldsymbol{\Phi}(z)$ and $\boldsymbol{\Psi}(z)$. The complex representation is therefore very advantageous because all tools of the complex analysis, such as power and Laurent series expansions, conformal mapping techniques or Cauchy's integral formula, can be used in problem solving. An important area of application is, for example, linear fracture mechanics for which the method of complex stress functions is the theoretical basis. Here the holomorphic functions $\boldsymbol{\Phi}(z)$ and $\boldsymbol{\Psi}(z)$ are approximated, for instance, by Laurent series expansions whereby analytic solutions to the differential equation in the near field of a crack tip can be constructed explicitly [17, 20]. These analytical near field representations have the specific property that the order of the singularity is precisely represented.

In this article a hypercomplex representation of the fundamental solution of three dimensional linear elasticity, also known as the Kelvin solution, is constructed. The Kelvin solution describes the displacements of a concentrated point load acting at the origin of an infinite body. On the basis of the superposition principle, solutions for forces groups or distributed forces can also be constructed. The derivation of the hypercomplex version of the Kelvin solution is based on a recent generalization of the Kolosov-Muskhelishvili formulae to $\mathbb{R}^3$ using the algebra $\mathbb{H}$ of real quaternions [8, 21]. For a detailed survey to this topic see [5, 21] and the references therein.

For this purpose, the article is structured as follows: Section 2 introduces the necessary foundations and notations regarding the non-commutative algebra $\mathbb{H}$. Then, Sect. 3 gives a brief overview of the derivation of the generalized Kolosov-Muskhelishvili formulae in $\mathbb{H}$. In Sect. 4, a new class of monogenic functions with (logarithmic) line singularities is studied and an associated two step recurrence formula is proved. In addition, the relationship of the constructed monogenic function system to the Cauchy-kernel function is established. These results are finally used in Sect. 5 to obtain the main result of the article that is a hypercomplex representation of the fundamental solution of the linear elasticity theory in $\mathbb{R}^3$.

2 Preliminaries and Notations

In order to generalize the complex setting to dimension three we will work in the algebra of real quaternions denoted by $\mathbb{H}$. Let $\{\mathbf{e}_0, \mathbf{e}_1, \mathbf{e}_2, \mathbf{e}_3\}$ be the standard basis subjected to the multiplication rules

$$\begin{aligned} \mathbf{e}_i\mathbf{e}_j + \mathbf{e}_j\mathbf{e}_i &= -2\delta_{ij}\,\mathbf{e}_0, \;\; i,j = 1,2,3, \\ \mathbf{e}_1\mathbf{e}_2 = \mathbf{e}_3, \;\; \mathbf{e}_0\mathbf{e}_i &= \mathbf{e}_i\mathbf{e}_0 = \mathbf{e}_i, \;\; i = 0,1,2,3. \end{aligned}$$

The real vector space $\mathbb{R}^4$ will be embedded in $\mathbb{H}$ by identifying the element $\boldsymbol{a} = [a_0, a_1, a_2, a_3]^{\mathrm{T}} \in \mathbb{R}^4$ with the quaternion $\boldsymbol{a} = a_0 + a_1\mathbf{e}_1 + a_2\mathbf{e}_2 + a_3\mathbf{e}_3$, $a_i \in \mathbb{R}$, $i = 0, 1, 2, 3$, where $\mathbf{e}_0 = [1, 0, 0, 0]^{\mathrm{T}}$ is the multiplicative unit element of the algebra $\mathbb{H}$. Further, we denote by

(i) $\mathbf{Sc}(\boldsymbol{a}) = a_0$ the scalar part, $\mathbf{Vec}(\boldsymbol{a}) = \underline{\boldsymbol{a}} = \sum_{i=1}^{3} a_i \mathbf{e}_i$ the vector part of $\boldsymbol{a}$,
(ii) $\overline{\boldsymbol{a}} = a_0 - \underline{\boldsymbol{a}}$ the conjugate of $\boldsymbol{a}$,
(iii) $\widehat{\boldsymbol{a}} = -\mathbf{e}_3 \boldsymbol{a} \mathbf{e}_3$ the $\mathbf{e}_3$-involution of $\boldsymbol{a}$,
(iv) $|\boldsymbol{a}| = \sqrt{\boldsymbol{a}\,\overline{\boldsymbol{a}}}$ the norm of $\boldsymbol{a}$,
(v) $\boldsymbol{a}^{-1} = \frac{\overline{\boldsymbol{a}}}{|\boldsymbol{a}|^2}$, $\boldsymbol{a} \neq 0$ the inverse of $\boldsymbol{a}$.

Now, let us consider the subset $\mathcal{A} := \operatorname{span}_{\mathbb{R}} \{1, \mathbf{e}_1, \mathbf{e}_2\}$. Here it is important to point out that $\mathcal{A}$ is only a real vector space but not a sub-algebra of $\mathbb{H}$. The real vector space $\mathbb{R}^3$ will be embedded in $\mathcal{A}$ by the identification of $\boldsymbol{x} = [x_0, x_1, x_2]^{\mathrm{T}} \in \mathbb{R}^3$ with the *reduced quaternion*

$$\boldsymbol{x} = x_0 + \mathbf{e}_1 \zeta \in \mathcal{A} \quad \text{with} \quad \zeta = x_1 - \mathbf{e}_3 x_2 .$$

As a consequence, the symbol $\boldsymbol{x}$ is often used to represent a point in $\mathbb{R}^3$ as well as to represent the corresponding reduced quaternion.

Let now Ω be an open subset of $\mathbb{R}^3$ with a piecewise smooth boundary. An $\mathbb{H}$-valued function is a mapping

$$\boldsymbol{f} : \Omega \longrightarrow \mathbb{H} \quad \text{such that} \quad \boldsymbol{f}(\boldsymbol{x}) = \sum_{i=0}^{3} f^i(\boldsymbol{x})\, \mathbf{e}_i \,, \ \boldsymbol{x} \in \Omega \,.$$

The coordinates $f^i(\boldsymbol{x})$ are real-valued functions defined in Ω, i.e., $f^i(\boldsymbol{x}) : \Omega \longrightarrow \mathbb{R}, i = 0, 1, 2, 3$. Continuity, differentiability or integrability of $\boldsymbol{f}$ are defined coordinate-wise. Note that the function $\boldsymbol{f}(\boldsymbol{x})$ is defined in $\mathbb{R}^3$ but taking values in $\mathbb{H} \cong \mathbb{R}^4$. Due to the underlying non commutative algebra all functions considered in the following will be considered in the right $\mathbb{H}$-linear Hilbert space of square-integrable $\mathbb{H}$-valued functions denoted by $L^2(\Omega; \mathbb{H})$. For a detailed discussion of the function spaces and the corresponding inner product, see e.g. [11].

In complex function theory the Cauchy-Riemann operator and its adjoint operator play an important role. In the introduced hypercomplex framework of $\mathbb{H}$-valued functions, the operator

$$\overline{\boldsymbol{\partial}} := \frac{\partial}{\partial x_0} + 2\mathbf{e}_1 \frac{\partial}{\partial \overline{\zeta}} \;=\; \frac{\partial}{\partial x_0} + \mathbf{e}_1 \frac{\partial}{\partial x_1} + \mathbf{e}_2 \frac{\partial}{\partial x_2} \tag{2.1}$$

is called *generalized Cauchy-Riemann operator*. The corresponding *adjoint generalized Cauchy-Riemann operator* is defined by

$$\boldsymbol{\partial} := \frac{\partial}{\partial x_0} - 2\mathbf{e}_1 \frac{\partial}{\partial \overline{\zeta}} \;=\; \frac{\partial}{\partial x_0} - \mathbf{e}_1 \frac{\partial}{\partial x_1} - \mathbf{e}_2 \frac{\partial}{\partial x_2} . \tag{2.2}$$

At this point, it is emphasized that throughout this article the introduced differential operators are considered as operators acting from the left and analogously denoted as in the complex analysis (see, e.g. [11]) which is vice versa to the originally introduced notation in Clifford analysis (see, e.g. [9]). This leads to the following definitions:

Definition 2.1 A function $\boldsymbol{f} \in C^1(\Omega; \mathbb{H})$ is called *monogenic* in $\Omega \subset \mathbb{R}^3$ if

$$\overline{\boldsymbol{\partial}}\boldsymbol{f} = 0 \text{ in } \Omega \text{ (or equivalently } \boldsymbol{f} \in \ker \overline{\boldsymbol{\partial}} \text{ in } \Omega). \tag{2.3}$$

Conversely, a function $\boldsymbol{g} \in C^1(\Omega; \mathbb{H})$ is called *anti-monogenic* in $\Omega \subset \mathbb{R}^3$ if

$$\boldsymbol{\partial}\boldsymbol{g} = 0 \text{ in } \Omega \text{ (or equivalently } \boldsymbol{g} \in \ker \boldsymbol{\partial} \text{ in } \Omega).$$

In view of the upcoming calculations we will often use the following component form

$$\boldsymbol{f} = \boldsymbol{f}^{03} + \mathbf{e}_1 \boldsymbol{f}^{12} := \left(f^0 + \mathbf{e}_3 f^3\right) + \mathbf{e}_1 \left(f^1 - \mathbf{e}_3 f^2\right)$$

of an $\mathbb{H}$-valued function $\boldsymbol{f} \in C^1(\Omega; \mathbb{H})$. Note that the multiplication of the components $\boldsymbol{f}^{03}, \boldsymbol{f}^{12}$ commutes with $\mathbf{e}_0, \mathbf{e}_3$ and anti-commutes with $\mathbf{e}_1, \mathbf{e}_2$. Accordingly, for $ij = \{03, 12\}$ we have

$$\mathbf{e}_k \boldsymbol{f}^{ij} = \begin{cases} \boldsymbol{f}^{ij} \mathbf{e}_k : k = 0, 3, \\ \overline{\boldsymbol{f}^{ij}} \mathbf{e}_k : k = 1, 2. \end{cases}$$

Applying now the compact formulation of the differential operator (2.1) to the component form of $\boldsymbol{f}$ yields an equivalent definition of monogenicity (2.3) given by the system

$$\boldsymbol{f} \in \ker \overline{\boldsymbol{\partial}} \;\Leftrightarrow\; \begin{cases} \dfrac{\partial \boldsymbol{f}^{03}}{\partial x_0} - 2\dfrac{\partial \boldsymbol{f}^{12}}{\partial \zeta} = 0, \\ \dfrac{\partial \boldsymbol{f}^{12}}{\partial x_0} + 2\dfrac{\partial \boldsymbol{f}^{03}}{\partial \overline{\zeta}} = 0. \end{cases} \tag{2.4}$$

Based on the last representation one can conclude a significant connection between monogenic and anti-monogenic functions.

Corollary 2.2 *Let* $\boldsymbol{f} = \sum_{i=0}^{3} f^i \mathbf{e}_i \in C^1(\Omega; \mathbb{H})$ *be a monogenic function in* $\Omega \subset \mathbb{R}^3$. *The* $\mathbf{e}_3$*-involution of the function*

$$\widehat{\boldsymbol{f}} := f^0 - f^1 \mathbf{e}_1 - f^2 \mathbf{e}_2 + f^3 \mathbf{e}_3$$

defines an anti-monogenic function in Ω.

Proof Let $f = f^{03} + \mathbf{e}_1 f^{12} \in \ker \overline{\partial}$, we have $\widehat{f} = f^{03} - \mathbf{e}_1 f^{12}$. Applying (2.2) and using the relations (2.4) yields

$$\partial \widehat{f} = \left(\frac{\partial f^{03}}{\partial x_0} - 2\frac{\partial f^{12}}{\partial \zeta} \right) - \mathbf{e}_1 \left(\frac{\partial f^{12}}{\partial x_0} + 2\frac{\partial f^{03}}{\partial \overline{\zeta}} \right) = 0.$$

□

Here, it should be emphasized that in the complex one-dimensional case the conjugation of a holomorphic function $f \in C^1(\Omega; \mathbb{C})$ gives directly the corresponding anti-holomorphic function $\widehat{f}$ and thus $\overline{f} \equiv \widehat{f}$. For $\mathbb{H}$-valued monogenic functions this property doesn't hold in general, as Corollary 2.2 shows. The subset of $\mathcal{A}$-valued functions is an exception to this, which can easily be seen from the foregoing calculations.

Furthermore, we need the concept of the hypercomplex derivative (see, e.g. the first works [10, 15] or for a survey [14]). The main result of [10] is summarized in the following definition.

Definition 2.3 (Hypercomplex Derivative) Let $f \in C^1(\Omega; \mathbb{H})$ be a continuous, real-differentiable function and monogenic in Ω. The expression $\partial_x f := \frac{1}{2}\partial f$ is called *hypercomplex derivative* of f in Ω.

As a consequence of Definition 2.3, we introduce a special subset of monogenic functions characterized by vanishing first derivatives.

Definition 2.4 (Monogenic Constant) A C^1-function belonging to $\ker \partial_x \cap \ker \overline{\partial}$ is called *monogenic constant.*

Finally, we state the definition of a monogenic primitive.

Definition 2.5 (Monogenic Primitive) A function $F \in C^1(\Omega, \mathbb{H})$ is called monogenic primitive of a monogenic function f with respect to the hypercomplex derivative ∂_x, if

$$F \in \ker \overline{\partial} \quad \text{and} \quad \partial_x F = f.$$

3 Generalized Kolosov-Muskhelishvili Formulae in $\mathbb{H}$

In this section a recent approach for the generalization of the well known Kolosov-Muskhelishvili formulae of plane linear elasticity to dimension three is briefly summarized. Particular emphasis is placed on the structural analogies between the complex and the hypercomplex representation.

3.1 General Solution of Papkovic-Neuber

In the linear theory of elasticity the equilibrium equation for an isotropic, homogeneous body in terms of the displacements (see, e.g. [4]) is given by the so-called *Lamé-Navier equation*

$$\mu\left(\Delta \underline{u} + \frac{1}{1-2\nu}\,\mathrm{grad}\,\mathrm{div}\,\underline{u}\right) = -\underline{F}, \tag{3.1}$$

where $\underline{u} = [u_0, u_1, u_2]^{\mathrm{T}}$ is the displacement field, $\underline{F}$ the vector of body forces and $\mu, \nu \in \mathbb{R}$ denote the material constants shear modulus and Poisson's ratio. Considering now the homogenous part of Eq. (3.1), the general solution of Papkovic [19] and Neuber [18] is given by

$$\left.\begin{aligned} 2\mu\, u_0 &= 4(1-\nu) f_0 - \frac{\partial G}{\partial x_0},\\ 2\mu\, u_1 &= 4(1-\nu) f_1 - \frac{\partial G}{\partial x_1},\\ 2\mu\, u_2 &= 4(1-\nu) f_2 - \frac{\partial G}{\partial x_2}. \end{aligned}\right\} \tag{3.2}$$

Here, $G \in C^4(\Omega;\mathbb{R})$ defines a bi-harmonic function

$$G = x_0 f_0 + x_1 f_1 + x_2 f_2 + h_0 \in \ker \Delta^2,$$

where $f_i \in \ker\Delta,\ i = 0, 1, 2$ and $h_0 \in \ker\Delta$ are arbitrary harmonic functions in $C^2(\Omega;\mathbb{R})$. Now, we reformulate the Papkovic-Neuber solution using the hypercomplex notation from the previous section. For this purpose, we denote the $\mathbb{H}$-valued representation of the displacement field with $\boldsymbol{u}^\star = u_0 + u_1\mathbf{e}_1 + u_2\mathbf{e}_2$ and the biharmonic function G with

$$G = \frac{1}{2}\left(\overline{\boldsymbol{x}}\,\mathfrak{f} + \overline{\mathfrak{f}}\,\boldsymbol{x}\right) + h_0,$$

where $\mathfrak{f} = f_0 + f_1\mathbf{e}_1 + f_2\mathbf{e}_2 \in \ker\Delta$. Consequently, the classical Papkovic-Neuber solution (3.2) reads in quaternionic algebra equivalently

$$2\mu\,\boldsymbol{u}^\star = 4(1-\nu)\,\mathfrak{f} - \frac{1}{2}\overline{\boldsymbol{\partial}}\left(\overline{\boldsymbol{x}}\,\mathfrak{f} + \overline{\mathfrak{f}}\,\boldsymbol{x} + 2h_0\right). \tag{3.3}$$

3.2 Kolosov-Muskhelishvili Formulae for the Displacements and Stresses

The following three dimensional generalizations of the classical Kolosov-Muskhelishvili formulae (1.1) for domains Ω which are normal w.r.t. the

x_0-direction were derived in detail in [8, 21]. Here the main idea of the derivation was that the harmonic function $\mathfrak{f} = f_0 + f_1\mathbf{e}_1 + f_2\mathbf{e}_2$ in Eq. (3.3) can be decomposed (see [12] or more recently [1, 2]) into a monogenic function $\boldsymbol{\Phi}$ and an anti-monogenic function $\boldsymbol{\Theta}$, such that $\mathfrak{f} = \boldsymbol{\Phi} + \boldsymbol{\Theta}$. Note, that even if in the decomposition the function $\mathfrak{f}$ is $\mathcal{A}$-valued the functions $\boldsymbol{\Phi}$ and $\boldsymbol{\Theta}$ can be $\mathbb{H}$-valued.

Applying the decomposition in the hypercomplex Papkovic-Neuber representation (3.3) one obtains the *generalized Kolosov-Muskhelishvili formula* for the extended displacement field $\boldsymbol{u} := \boldsymbol{u}^{\star} + \chi\mathbf{e}_3$ given by

$$2\mu\boldsymbol{u} = 4(1-\nu)\,\boldsymbol{\Phi} - \frac{1}{2}\overline{\boldsymbol{\partial}}\left(\overline{\boldsymbol{x}}\,\boldsymbol{\Phi} + \overline{\boldsymbol{\Phi}}\,\boldsymbol{x}\right) - \widehat{\boldsymbol{\Psi}} \tag{3.4}$$

in terms of a monogenic function $\boldsymbol{\Phi} \in \ker\overline{\boldsymbol{\partial}} \perp (\ker\boldsymbol{\partial} \cap \ker\overline{\boldsymbol{\partial}})$ orthogonal to the subset of monogenic constants and an anti-monogenic function $\widehat{\boldsymbol{\Psi}} \in \ker\boldsymbol{\partial}$. The function χ defines a harmonic function

$$2\mu\chi = \left[4(1-\nu)\,\boldsymbol{\Phi} - \widehat{\boldsymbol{\Psi}}\right]_{\mathbf{e}_3} \in C^1(\Omega;\mathbb{R}),$$

where $[f]_{\mathbf{e}_3}$ denotes the $\mathbf{e}_3$-part of $\boldsymbol{f}$. Based on the extended displacement field one can also derive generalized Kolosov-Muskhelishvili formulae for the stresses. Analogously to the complex case we first calculate the trace of the stress tensor and obtain the 1st *generalized Kolosov-Muskhelishvili formulae for the stresses* given by

$$\sigma_{00} + \sigma_{11} + \sigma_{22} = (1+\nu)\left(\boldsymbol{\partial}\boldsymbol{\Phi} + \overline{\boldsymbol{\partial}\boldsymbol{\Phi}}\right) = 2(1+\nu)\,\mathbf{Sc}\left(\boldsymbol{\partial}\boldsymbol{\Phi}\right). \tag{3.5}$$

Using the following abbreviations

$$\mathfrak{S}_j(\boldsymbol{u}) = \begin{cases} \sigma_{00} - \sigma_{11} - \sigma_{22} + 2\tau_{01}\mathbf{e}_1 + 2\tau_{02}\mathbf{e}_2 + 2\mu\dfrac{\partial\chi}{\partial x_0}\mathbf{e}_3 : j = 0, \\ -\sigma_{00} + \sigma_{11} - \sigma_{22} - 2\tau_{01}\mathbf{e}_1 + 2\mu\dfrac{\partial\chi}{\partial x_1}\mathbf{e}_2 - 2\tau_{12}\mathbf{e}_3 : j = 1, \\ -\sigma_{00} - \sigma_{11} + \sigma_{22} - 2\mu\dfrac{\partial\chi}{\partial x_2}\mathbf{e}_1 - 2\tau_{02}\mathbf{e}_2 + 2\tau_{12}\mathbf{e}_3 : j = 2 \end{cases}$$

the remaining *generalized Kolosov-Muskhelishvili formulae* for the stresses defined in terms of the extended displacement field are

$$\mathfrak{S}_j(\boldsymbol{u}) = 2\mu\left(\frac{\nu-1}{2(1-2\nu)}\left(\boldsymbol{\partial}\boldsymbol{u} + \overline{\boldsymbol{\partial}\boldsymbol{u}}\right) + \frac{1}{2}\left(\boldsymbol{\partial}\boldsymbol{u} + \overline{\mathbf{e}}_j\overline{\boldsymbol{u}}\overline{\boldsymbol{\partial}}\mathbf{e}_j\right) + \mathbf{e}_j\overline{\boldsymbol{\partial}}\mathbf{e}_j\boldsymbol{u}\right). \tag{3.6}$$

For a detailed derivation of the formulae (3.4)–(3.6) see [8, 21].

The generalized Kolosov-Muskhelishvili formulae can now serve as a basis for solving boundary value problems of linear elasticity theory in $\mathbb{R}^3$ by means of hypercomplex methods. As in the complex theory, efficient series expansions of monogenic functions are also available in the hypercomplex framework. In [5, 7] an orthogonal basis of monogenic polynomials is studied and generalized *Taylor* and *Fourier series* in $\mathbb{H}$ are defined. These series expansions, together with the generalized Kolosov-Muskhelishvili formula for the displacements (3.4), are then applied in [8] to construct a complete system of polynomial solutions to the Lamé-Navier equation (3.1). Furthermore, a generalized *Laurent series* is defined in [6] on the basis of an complete orthogonal system of outer spherical monogenic functions. These results generalize most of the properties of the complex series expansions in terms of the holomorphic z-powers to the spatial case.

4 Monogenic Primitives of the Cauchy-Kernel Function

As the starting point, we consider the *Legendre differential equation*

$$(1-t^2)\,\frac{\mathrm{d}^2y}{\mathrm{d}t^2} - 2t\,\frac{\mathrm{d}y}{\mathrm{d}t} + n(n+1)y = 0, \quad n > 0,\ |t| < 1\,. \tag{4.1}$$

It is well known that the general solution of the second-order ordinary differential equation (4.1) is given by

$$y = A\,P_n(t) + B\,Q_n(t),$$

where $P_n(t)$ and $Q_n(t)$ are the Legendre functions of the 1st and 2nd kind, respectively. In previous works [6–8] it was shown that the Legendre functions $P_n(t)$ as well as the associated Legendre functions $P_n^m(t)$ of the 1st kind play a crucial role in defining orthogonal Appell basis of inner and outer spherical monogenics. Here, we will work with the second linearly independent solutions $Q_n(t)$ of the Legendre differential equation and construct a system of special monogenic functions having a logarithmic singularity along the whole x_0-axis. In [16], a system of prolate spheroidal monogenic functions was constructed. These functions have a logarithmic point singularity in the origin and asymptotically behave like the classical outer spherical monogenics. For these reasons, the prolate spheroidal monogenic functions and their properties differ considerably from the functions constructed here.

In the end it is shown that the functions studied in this work can be interpreted as monogenic primitives of the Cauchy-kernel function in $\mathbb{R}^3$. The term *monogenic primitives* is used in this context to describe functions that can be constructed by the k-fold application ($k \in \mathbb{N}$) of a monogenic primitivation operator to the Cauchy-kernel function.

4.1 Construction of Monogenic Functions with a Logarithmic Line Singularity

The Legendre functions $Q_n(t)$ of the 2nd kind can be defined, see for example [3], by the recurrence relation

$$(n+1)\,Q_{n+1}(t) = (2n+1)\,t\,Q_n(t) - n\,Q_{n-1}(t), \quad n \geq 1 \tag{4.2}$$

with

$$Q_0(t) = \frac{1}{2}\ln\frac{1+t}{1-t} \qquad \text{and} \qquad Q_1(t) = t\,Q_0(t) - 1\,.$$

Due to the logarithmic term in $Q_0(t)$ the functions $Q_n(t)$ have infinite discontinuities at $t = \pm 1$. Therefore, we consider in the following the functions $Q_n(t)$ on the interval $|t| < 1$. For the upcoming calculations, we need also the following recurrence relations, [3]. For $n \geq 1$, the functions $Q_n(t)$ satisfy the relations

$$Q'_{n+1}(t) - 2t\,Q'_n(t) + Q'_{n-1}(t) - Q_n(t) = 0, \tag{4.3}$$

$$(1-t^2)\,Q'_n(t) - n\big(Q_{n-1}(t) - t\,Q_n(t)\big) = 0, \tag{4.4}$$

$$t\,Q'_n(t) - Q'_{n-1}(t) - n\,Q_n(t) = 0. \tag{4.5}$$

Now, let us define $H_n(\boldsymbol{x}) := |\boldsymbol{x}|^n Q_n\left(\frac{x_0}{|\boldsymbol{x}|}\right)$, $n \in \mathbb{N}_0$. It is known (see, e.g. [4]) that the functions $H_n(\boldsymbol{x})$ are harmonic. Hence, applying the known factorization of the Laplace operator by the generalized Cauchy-Riemann operator (2.1) and its adjoint operator (2.2), $\Delta = \overline{\boldsymbol{\partial}}\boldsymbol{\partial}$, for each $n \geq -1$ we get a monogenic function through

$$\boldsymbol{W}_n(\boldsymbol{x}) := \boldsymbol{\partial} H_{n+1}(\boldsymbol{x}) = \boldsymbol{\partial}\left(|\boldsymbol{x}|^{n+1} Q_{n+1}\left(\frac{x_0}{|\boldsymbol{x}|}\right)\right) \in \ker \overline{\boldsymbol{\partial}}\,. \tag{4.6}$$

Next, we evaluate the right hand side of (4.6) to obtain a differential operator free representation of the functions $\boldsymbol{W}_n(\boldsymbol{x})$. Note that subsequently in the context of monogenic functions we often use the substitution $t = \frac{x_0}{|\boldsymbol{x}|}$ together with $\boldsymbol{x}$ and $\boldsymbol{\zeta}$ to simplify the notation. Thus, we have to compute

$$\begin{aligned}
\boldsymbol{W}_n(\boldsymbol{x}) &= \boldsymbol{\partial}\left(|\boldsymbol{x}|^{n+1} Q_{n+1}(t)\right) \\
&= \left(\frac{\partial}{\partial x_0}|\boldsymbol{x}|^{n+1} Q_{n+1}(t)\right) - 2\mathbf{e}_1\left(\frac{\partial}{\partial \overline{\boldsymbol{\zeta}}}|\boldsymbol{x}|^{n+1} Q_{n+1}(t)\right).
\end{aligned}$$

By a straightforward calculation, for the first term in brackets we get

$$\frac{\partial}{\partial x_0}|\boldsymbol{x}|^{n+1} Q_{n+1}(t) = (n+1)\,x_0\,|\boldsymbol{x}|^{n-1}\,Q_{n+1}(t) + \boldsymbol{\zeta}\overline{\boldsymbol{\zeta}}\,|\boldsymbol{x}|^{n-2}\,Q'_{n+1}(t)$$

and for the second term in brackets

$$\frac{\partial}{\partial \bar{\zeta}} |\boldsymbol{x}|^{n+1} Q_{n+1}(t) = \frac{1}{2}\Big[(n+1)\, \boldsymbol{\zeta}\, |\boldsymbol{x}|^{n-1}\, Q_{n+1}(t) - x_0 \boldsymbol{\zeta}\, |\boldsymbol{x}|^{n-2}\, Q'_{n+1}(t)\Big]\,.$$

Accordingly, for each $n \geq -1$ we obtain

$$\boldsymbol{W}_n(\boldsymbol{x}) = (n+1)\, \bar{\boldsymbol{x}}\, |\boldsymbol{x}|^{n-1}\, Q_{n+1}(t) \;+\; \mathbf{e}_1 \boldsymbol{\zeta} \bar{\boldsymbol{x}}\, |\boldsymbol{x}|^{n-2}\, Q'_{n+1}(t)\,. \tag{4.7}$$

Finally, we also give a representation of (4.7) in spherical coordinates. We introduce spherical coordinates by

$$x_0 = r\cos\theta, \quad x_1 = r\sin\theta\cos\varphi, \quad x_2 = r\sin\theta\sin\varphi,$$

where $0 < r < \infty$, $0 < \theta \leq \pi$, $0 < \varphi \leq 2\pi$. Each $\boldsymbol{x} \in \mathbb{R}^3 \setminus \{\mathbf{0}\}$ admits a unique representation $\boldsymbol{x} = r\boldsymbol{\omega}$, where $\boldsymbol{\omega} = \omega_0 + \omega_1 \mathbf{e}_1 + \omega_2 \mathbf{e}_2$, with $\omega_j = \frac{x_j}{r}$ $(j = 0, 1, 2)$ and $|\boldsymbol{\omega}| = 1$. With $|\boldsymbol{x}| = r$ and $t = \cos\theta$ we get

$$\begin{aligned} \boldsymbol{W}_n(r, \boldsymbol{\omega}) = r^n\Big[\; & (n+1)\cos\theta\, Q_n(\cos\theta) + \sin^2\theta\, Q'_{n+1}(\cos\theta) \\ & + \Big(\cos\theta\sin\theta\, Q'_{n+1}(\cos\theta) - (n+1)\sin\theta\, Q_{n+1}(\cos\theta)\Big) \\ & \times \Big(\mathbf{e}_1\cos\varphi + \mathbf{e}_2\sin\varphi\Big)\Big]. \end{aligned}$$

4.2 Recurrence Relations

From a computational point of view it is always convenient to have recurrence relations since the whole system of functions up to a fixed integer n can be constructed out of initial functions. This means that, apart from the initial functions of the recursion, no additional functions or function libraries need to be implemented. The main result of this section is stated in the following theorem.

Theorem 4.1 *For each $n \in \mathbb{N}$, the monogenic functions $\boldsymbol{W}_n(\boldsymbol{x})$, explicitly defined in Eq. (4.7), satisfy the two-step recurrence formula*

$$\boldsymbol{W}_n(\boldsymbol{x}) \;=\; \left(2x_0 + \mathbf{e}_1 \frac{\zeta}{n}\right) \boldsymbol{W}_{n-1}(\boldsymbol{x}) - \boldsymbol{x}\bar{\boldsymbol{x}}\, \boldsymbol{W}_{n-2}(\boldsymbol{x}) \tag{4.8}$$

with the initial functions

$$\boldsymbol{W}_{-1}(\boldsymbol{x}) = \frac{\boldsymbol{\zeta}\bar{\boldsymbol{\zeta}} + \mathbf{e}_1 x_0 \boldsymbol{\zeta}}{|\boldsymbol{x}|\boldsymbol{\zeta}\bar{\boldsymbol{\zeta}}} \quad \textit{and} \quad \boldsymbol{W}_0(\boldsymbol{x}) = \frac{1}{2}\ln\left(\frac{|\boldsymbol{x}| + x_0}{|\boldsymbol{x}| - x_0}\right) + \frac{\mathbf{e}_1 |\boldsymbol{x}| \boldsymbol{\zeta}}{\boldsymbol{\zeta}\bar{\boldsymbol{\zeta}}}\,.$$

Proof First of all, we write the functions $\boldsymbol{W}_n(\boldsymbol{x})$ using Eq. (4.7) in component form

$$\begin{aligned}\boldsymbol{W}_n(\boldsymbol{x}) &= \boldsymbol{W}_n^{03}(\boldsymbol{x}) + \mathbf{e}_1\, \boldsymbol{W}_n^{12}(\boldsymbol{x})\\ &= (n+1)x_0|\boldsymbol{x}|^{n-1}Q_{n+1}(t) + \zeta\overline{\zeta}\, |\boldsymbol{x}|^{n-2}Q'_{n+1}(t)\\ &\quad + \mathbf{e}_1\Big[x_0\zeta\, |\boldsymbol{x}|^{n-2}Q'_{n+1}(t) - (n+1)\zeta|\boldsymbol{x}|^{n-1}Q_{n+1}(t)\Big].\end{aligned}$$

In the last equation we use the identities

$$\frac{x_0}{|\boldsymbol{x}|} = t \qquad \text{and} \qquad \frac{\zeta\overline{\zeta}}{|\boldsymbol{x}|^2} = \frac{|\boldsymbol{x}|^2 - x_0^2}{|\boldsymbol{x}|^2} = 1 - t^2,$$

and we get

$$\begin{aligned}\boldsymbol{W}_n(\boldsymbol{x}) = |\boldsymbol{x}|^n \Big[&(n+1)\, t\, Q_{n+1}(t) + (1-t^2)Q'_{n+1}(t)\\ &+ \mathbf{e}_1 \frac{\zeta}{|\boldsymbol{x}|}\Big(t\, Q'_{n+1}(t) - (n+1)\, Q_{n+1}(t)\Big)\Big].\end{aligned}$$

Now we apply the recurrence relations (4.4) and (4.5) to $\boldsymbol{W}_n^{03}(\boldsymbol{x})$ and $\boldsymbol{W}_n^{12}(\boldsymbol{x})$, respectively. For each $n \geq 0$, we finally get

$$\boldsymbol{W}_n(\boldsymbol{x}) = |\boldsymbol{x}|^n \left[(n+1)\, Q_n(t) + \mathbf{e}_1 \frac{\zeta}{|\boldsymbol{x}|}\, Q'_n(t) \right]. \tag{4.9}$$

For the proof of the recurrence relation (4.8) we first consider the case $n = 1$. With (4.9) and (4.2) we obtain for the left hand side of (4.8) the relation

$$\boldsymbol{W}_1(\boldsymbol{x}) = 2\,|\boldsymbol{x}| \left(\frac{1}{2}\, t \ln\left(\frac{1+t}{1-t}\right) - 1\right) + \mathbf{e}_1\zeta \left(\frac{1}{2} \ln\left(\frac{1+t}{1-t}\right) + \frac{x_0}{\zeta\overline{\zeta}}\right).$$

For the right hand side of (4.8) we get equivalently

$$\begin{aligned}&(2x_0 + \mathbf{e}_1\boldsymbol{\zeta})\ \boldsymbol{W}_0(\boldsymbol{x}) - \boldsymbol{x}\overline{\boldsymbol{x}}\, \boldsymbol{W}_{-1}(\boldsymbol{x})\\ &= (2x_0 + \mathbf{e}_1\zeta) \left(\frac{1}{2} \ln\left(\frac{1+t}{1-t}\right) + \frac{\mathbf{e}_1|\boldsymbol{x}|\zeta}{\zeta\overline{\zeta}}\right) - |\boldsymbol{x}|^2\, \frac{\zeta\overline{\zeta} + \mathbf{e}_1 x_0\zeta}{|\boldsymbol{x}|\zeta\overline{\zeta}} \equiv W_1(\boldsymbol{x}).\end{aligned}$$

Let us now consider the case $n > 1$. Representing the functions $\boldsymbol{W}_n(\boldsymbol{x})$ of (4.8) in component form, it is equivalent to prove that

$$\begin{aligned}&n\, \boldsymbol{W}_n(\boldsymbol{x}) = (2n\, x_0 + \mathbf{e}_1\zeta)\, \boldsymbol{W}_{n-1}(\boldsymbol{x}) - n\, |\boldsymbol{x}|^2\, \boldsymbol{W}_{n-2}(\boldsymbol{x})\\ \Leftrightarrow &\begin{cases} \text{(i) } n\, \boldsymbol{W}_n^{03}(\boldsymbol{x}) = 2n\, x_0\, \boldsymbol{W}_{n-1}^{03}(\boldsymbol{x}) - \overline{\zeta}\, \boldsymbol{W}_{n-1}^{12}(\boldsymbol{x}) - n\, |\boldsymbol{x}|^2\, \boldsymbol{W}_{n-2}^{03}(\boldsymbol{x}),\\ \text{(ii) } n\, \boldsymbol{W}_n^{12}(\boldsymbol{x}) = 2n\, x_0\, \boldsymbol{W}_{n-1}^{12}(\boldsymbol{x}) + \zeta\, \boldsymbol{W}_{n-1}^{03}(\boldsymbol{x}) - n\, |\boldsymbol{x}|^2\, \boldsymbol{W}_{n-2}^{12}(\boldsymbol{x}).\end{cases}\end{aligned}$$

In order to prove (i), we substitute the components of $\mathbf{W}_n(\mathbf{x})$ based on the representation (4.9) and obtain

$$0 = |\mathbf{x}|^n \big[n(n+1) Q_n(t) \; - \; 2n^2 \, t \, Q_{n-1}(t) \\ + (1-t^2) \, Q'_{n-1}(t) \; + \; n(n-1) \, Q_{n-2}(t) \big] .$$

Applying in the last equation the recurrence relation (4.2), the term in brackets simplifies to the left hand side of recurrence relation (4.4)

$$0 = |\mathbf{x}|^n \Big[(n-1) \Big(Q_{n-2}(t) - t \, Q_{n-1}(t) \Big) - (1-t^2) Q'_{n-1}(t) \Big]$$

and thus proves (i). To verify (ii), we again replace the components of $\mathbf{W}_n(\mathbf{x})$ by the relation (4.9) and obtain by a straightforward calculation

$$0 = n \, |\mathbf{x}|^{n-1} \zeta \Big[Q'_n(t) - 2 \, t \, Q'_{n-1}(t) + Q'_{n-2}(t) - Q_{n-1}(t) \Big] .$$

The term in brackets is equal to the left hand side of recurrence relation (4.3) and consequently also proves (ii). □

4.3 The Connection with the Cauchy-Kernel Function

In this section, we show that the constructed system $\{\mathbf{W}_n(\mathbf{x})\}_{n \geq -1}$ of monogenic functions can be interpreted as monogenic primitives (see, Definition 2.5) of the Cauchy-kernel function in $\mathbb{R}^3$, given by

$$\mathbf{E}(\mathbf{x}) = \frac{1}{4\pi} \frac{\overline{\mathbf{x}}}{|\mathbf{x}|^3}, \quad \mathbf{x} \neq 0. \tag{4.10}$$

To this end, we first have to prove the following theorem.

Theorem 4.2 *For a fixed $n \geq 0$, the application of the hypercomplex derivative $\partial_{\mathbf{x}} = \frac{1}{2}\partial$ to the function $\mathbf{W}_n(\mathbf{x})$ yields*

$$\partial_{\mathbf{x}} \mathbf{W}_n(\mathbf{x}) \;=\; (n+1) \mathbf{W}_{n-1}(\mathbf{x}) .$$

Proof To proof the theorem, we consider first the case $n = 0$. A straightforward computation gives directly

$$\partial_{\mathbf{x}} \mathbf{W}_0(\mathbf{x}) = \partial_{\mathbf{x}} \left(\frac{1}{2} \ln \left(\frac{|\mathbf{x}| + x_0}{|\mathbf{x}| - x_0} \right) + \frac{\mathbf{e}_1 |\mathbf{x}| \zeta}{\zeta \overline{\zeta}} \right) = \frac{\zeta \overline{\zeta} + \mathbf{e}_1 x_0 \zeta}{|\mathbf{x}| \zeta \overline{\zeta}} \equiv \mathbf{W}_{-1}(\mathbf{x}).$$

For the case $n > 0$, we work again with the component representations of the function (4.9) as well as the differential operators (2.1) and (2.2), respectively. Thus, we have to calculate

$$\left(\frac{\partial}{\partial x_0} \pm 2\mathbf{e}_1 \frac{\partial}{\partial \overline{\zeta}}\right)\left(\mathbf{W}_n^{03}(\mathbf{x}) + \mathbf{e}_1 \mathbf{W}_n^{12}(\mathbf{x})\right)$$

$$= \frac{\partial}{\partial x_0}\mathbf{W}_n^{03}(\mathbf{x}) \mp 2\frac{\partial}{\partial \zeta}\mathbf{W}_n^{12}(\mathbf{x}) + \mathbf{e}_1\left[\frac{\partial}{\partial x_0}\mathbf{W}_n^{12}(\mathbf{x}) \pm 2\frac{\partial}{\partial \overline{\zeta}}\mathbf{W}_n^{03}(\mathbf{x})\right].$$

Using the Legendre differential equation (4.1) as well as the recurrence relations (4.4) and (4.5), by straightforward calculations we get

$$\begin{cases} \dfrac{\partial}{\partial x_0}\mathbf{W}_n^{03}(\mathbf{x}) = 2\dfrac{\partial}{\partial \zeta}\mathbf{W}_n^{12}(\mathbf{x}) = n(n+1)\,|\mathbf{x}|^{n-1}Q_{n-1}(t), \\ \dfrac{\partial}{\partial x_0}\mathbf{W}_n^{12}(\mathbf{x}) = -2\dfrac{\partial}{\partial \overline{\zeta}}\mathbf{W}_n^{03}(\mathbf{x}) = (n+1)\,\zeta\,|\mathbf{x}|^{n-2}Q'_{n-1}(t). \end{cases}$$

For each $n > 0$, we have

$$\left(\frac{\partial}{\partial x_0} \pm 2\mathbf{e}_1 \frac{\partial}{\partial \overline{\zeta}}\right)\mathbf{W}_n(\mathbf{x}) = (1 \mp 1)\,(n+1)|\mathbf{x}|^{n-1}\left[\,n\,Q_{n-1}(t) + \mathbf{e}_1\frac{\zeta}{|\mathbf{x}|}\,Q'_{n-1}(t)\,\right]$$

proving that

$$\boldsymbol{\partial}_x \mathbf{W}_n(\mathbf{x}) \;=\; (n+1)\mathbf{W}_{n-1}(\mathbf{x}) \quad \text{and} \quad \overline{\boldsymbol{\partial}}\,\mathbf{W}_n(\mathbf{x}) \;=\; 0.$$

□

From Theorem 4.2 we further conclude that for each $\mathbf{W}_n(\mathbf{x})$, $n \geq -1$, there exists a monogenic primitive

$$\mathbb{P}\,\mathbf{W}_n(\mathbf{x}) = \frac{1}{n+2}\mathbf{W}_{n+1}(\mathbf{x}) \in \ker\overline{\boldsymbol{\partial}} \quad \text{with} \quad \boldsymbol{\partial}_x\Big(\mathbb{P}\,\mathbf{W}_n(\mathbf{x})\Big) = \mathbf{W}_n(\mathbf{x}).$$

Finally, we calculate the hypercomplex derivative of the function $\mathbf{W}_{-1}(\mathbf{x})$ and we get

$$\boldsymbol{\partial}_x\,\mathbf{W}_{-1}(\mathbf{x}) = \boldsymbol{\partial}_x\left(\frac{\zeta\overline{\zeta} + x_0\mathbf{e}_1\zeta}{|\mathbf{x}|\zeta\overline{\zeta}}\right) = -\frac{\overline{\mathbf{x}}}{|\mathbf{x}|^3} \equiv -4\pi\,\mathbf{E}(\mathbf{x}),$$

which leads to the Cauchy-kernel function (4.10) except for a normalization factor.

5 The Kelvin Solution in $\mathbb{H}$

In this section, we study an important problem in linear elasticity, the so called *fundamental solution* or *Kelvin solution*, which describes the elastic displacements $\boldsymbol{u}^\star$ of a concentrated force $\boldsymbol{F}$ acting at the origin of an infinite body. The Kelvin solution is of fundamental importance as one can also construct solutions for force groups or distributed forces by using the principle of superposition.

In order to obtain a hypercomplex version of the Kelvin solution in the framework of the generalized Kolosov-Muskhelishvili formulae, we first consider the Kelvin solution formulated in terms of the Papkovich-Neuber potentials (3.2). Based on the representation (3.3), the Kelvin solution reads

$$2\mu\,\boldsymbol{u}^\star = 2\alpha\,\mathbf{g} - \frac{1}{2}\overline{\partial}\,(\overline{\boldsymbol{x}}\,\mathbf{g} + \overline{\mathbf{g}}\boldsymbol{x}),$$

where $\mathbf{g}$ denotes a vector valued harmonic function, given by

$$\mathbf{g}(\boldsymbol{x}) = \frac{\boldsymbol{F}}{4\pi\alpha\,|\boldsymbol{x}|} \in \ker\Delta \quad \text{and} \quad \boldsymbol{F} = F_0 + F_1\mathbf{e}_1 + F_2\mathbf{e}_2.$$

Applying now the results form the previous Sect. 4, we are able to decompose $\mathbf{g}$ into a monogenic function $\boldsymbol{\Phi}$ and an anti-monogenic function $\boldsymbol{\Theta}$ in the following way

$$\mathbf{g} = \boldsymbol{\Phi} + \boldsymbol{\Theta} = \frac{1}{2}\left(\boldsymbol{W}_{-1}(\boldsymbol{x}) + \widehat{\boldsymbol{W}_{-1}(\boldsymbol{x})}\right)\frac{\boldsymbol{F}}{4\pi\alpha},$$

where

$$\boldsymbol{\Phi}(\boldsymbol{x}) = \frac{\zeta\overline{\zeta} + \mathbf{e}_1 x_0\zeta}{2\,\zeta\overline{\zeta}\,|\boldsymbol{x}|}\frac{\boldsymbol{F}}{4\pi\alpha} \quad \text{and} \quad \boldsymbol{\Theta}(\boldsymbol{x}) = \frac{\zeta\overline{\zeta} - \mathbf{e}_1 x_0\zeta}{2\,\zeta\overline{\zeta}\,|\boldsymbol{x}|}\frac{\boldsymbol{F}}{4\pi\alpha}.$$

Let us here note the interesting fact that the function $\mathbf{g}$, which has a point singularity in the origin, is decomposed in the hypercomplex representation by two functions $\boldsymbol{\Phi}$ and $\boldsymbol{\Theta}$ with a line singularity on the entire real axis x_0.

Consequently, we obtain regarding the *generalized Kolosov-Muskhelishvili formula for the displacements* the following hypercomplex version of the classical Kelvin solution

$$2\mu\,\boldsymbol{u} = 2\alpha\,\boldsymbol{\Phi} - \frac{1}{2}\overline{\partial}\left(\overline{\boldsymbol{x}}\,\boldsymbol{\Phi} + \overline{\boldsymbol{\Phi}}\,\boldsymbol{x}\right) - \widehat{\boldsymbol{\Psi}}$$

where

$$\widehat{\boldsymbol{\Psi}} = \frac{1}{2}\overline{\partial}\left(\overline{\boldsymbol{x}}\,\boldsymbol{\Theta} + \overline{\boldsymbol{\Theta}}\,\boldsymbol{x}\right) - 2\alpha\,\boldsymbol{\Theta}.$$

6 Conclusions

In the article the fundamental solution of linear elasticity in $\mathbb{R}^3$ is represented by means of generalized Kolosov-Muskhelishvili formulae in the algebra of real quaternions $\mathbb{H}$. For this purpose, a new class of monogenic functions with logarithmic singularities is studied and a compact two step recurrence relation is proved. These functions extend the known orthogonal basis systems of outer and inner spherical monogenics, and can be characterized as monogenic primitives of the Cauchy-kernel function. The special set of monogenic functions with logarithmic line singularities could also play an important role in the derivation of generalized Kolosov-Muskhelishvili formulae for multiple connected domains or in applications in spatial linear fracture mechanics.

References

1. C. Álvarez-Peña, *Contragenic Functions and Appell Bases for Monogenic Functions of Three Variables*. Ph.D. thesis. Centro de Investigacion y de Estudios Avanzados del I.P.N., Mexico (2013)
2. C. Álvarez-Peña, R.M. Porter, Contragenic functions of three variables. Compl. Anal. Oper. Theory **8**, 409–427 (2014)
3. L.C. Andrews, *Special Functions of Mathematics for Engineers* (SPIE Optical Engineering Press, Bellingham; Oxford University Press, Oxford, 1998)
4. J.R. Barber, *Elasticity*. Solid Mechanics and Its Applications, vol. 172, 3rd rev. edn. (Springer, New York, 2010)
5. S. Bock, K. Gürlebeck, On a spatial generalization of the Kolosov-Muskhelishvili formulae. Math. Methods Appl. Sci. **32**, 223–240 (2009)
6. S. Bock, On a three dimensional analogue to the holomorphic z-powers: Laurent series expansions. Compl. Var. Elliptic Equ. **57**(12), 1271–1287 (2012)
7. S. Bock, On a three dimensional analogue to the holomorphic z-powers: power series and recurrence formulae. Compl. Var. Elliptic Equ. **57**(12), 1349–1370 (2012)
8. S. Bock, On monogenic series expansions with applications to linear elasticity. Adv. Appl. Clifford Algebr. **24**(4), 931–943 (2014)
9. F. Brackx, R. Delanghe, F. Sommen, Clifford Analysis. Pitman Research Notes Math. Ser. 76, Pitman, London etc. (1982)
10. K. Gürlebeck, H.R. Malonek, A hypercomplex derivative of monogenic functions in $\mathbb{R}^{n+1}$ and its applications. Compl. Var. **39**, 199–228 (1999)
11. K. Gürlebeck, K. Habetha, W. Sprößig, Holomorphic functions in the plane and n-dimensional space, in *A Birkhäuser Book* (2008), ISBN: 978-3-7643-8271-1
12. B. Klein Obbink, *On the Solutions of D^nD^mF*. Reports on Applied and Numerical Analysis (Eindhoven University of Technology, Department of Mathematics and Computing Science, 1993)
13. G.W. Kolosov, *Über einige Eigenschaften des ebenen Problems der Elastizitätstheorie*, Z. Math. Phys. **62**, 383–409 (1914)
14. M.E. Luna-Elizarrarás, M. Shapiro, A survey on the (hyper-) derivates in complex, quaternionic and Clifford analysis. Millan J. Math. **79**, 521–542 (2011)
15. H.R. Malonek, *Zum Holomorphiebegriff in höheren Dimensionen*, Habilitationsschrift. Pädagogische Hochschule Halle (1987)

16. J. Morais, M.H. Nguyen, K.I. Kou, On 3D orthogonal prolate spheroidal monogenics. Math. Methods Appl. Sci. **39**(4), 635–648 (2016)
17. N.I. Muskhelishvili, *Some Basic Problems of the Mathematical Theory of Elasticity*,(translated from the Russian by J.R.M. Radok.) (Noordhoff International Publishing, Leyden, 1977)
18. H. Neuber, *Ein neuer Ansatz zur Lösung räumlicher Probleme der Elastizitätstheorie; der Hohlkegel unter Einzellast als Beispiel*. Z. Angew. Math. Mech. **14**, 203–212 (1934)
19. P. Papkovic, *Solution générale des équations différentielles fondamentales de l'élasticité, exprimée par un vecteur et un scalaire harmonique* (Russisch), in Bull. Acad. Sc. Leningrad (1932), pp. 1425–1435
20. J.R. Rice, Mathematical analysis in the mechanics of fracture, in *Fracture, An Advanced Treatise*, ed. by H. Liebowitz. Mathematical Fundamentals, vol. 2 (Academic Press, New York, 1968), pp. 191–311
21. D. Weisz-Patrault, S. Bock, K. Gürlebeck, Three-dimensional elasticity based on quaternion-valued potentials. Int. J. Solids Struct. **51**(19), 3422–3430 (2014)

Viscoelastic Behavior of Periodontal Ligament: Stresses Relaxation at Translational Displacement of a Tooth Root

S. Bosiakov, G. Mikhasev, and S. Rogosin

Abstract Understanding of viscoelastic response of a periodontal membrane under the action of short-term and long-term loadings is important for many orthodontic problems. A new analytic model describing behavior of the viscoelastic periodontal ligament after the tooth root translational displacement based on Maxwell approach is suggested. In the model, a tooth root and alveolar bone are assumed to be a rigid bodies. The system of differential equations for the plane-strain state of the viscoelastic periodontal ligament is used as the governing one. The boundary conditions corresponding to the initial small displacement of the root and fixed outer surface of the periodontal ligament in the dental alveolus are utilized. A solution is found numerically for fractional viscoelasticity model assuming that the stress relaxation in the periodontal ligament after the continuing displacement of the tooth root occurs approximately within five hours. The character of stress distribution in the ligament over time caused by the tooth root translational displacement is evaluated. Effect of Poisson's ratio on the stresses in the viscoelastic periodontal ligament is considered. The obtained results can be used for simulation of the bone remodelling process during orthodontic treatment and for assessment of optimal conditions of the orthodontic load application.

Keywords Viscoelastic periodontal ligament • Translational displacement • Root of the tooth • Stress relaxation

Mathematics Subject Classification (2010) 74L15, 35Q92, 45E10

S. Bosiakov • G. Mikhasev • S. Rogosin (✉)
Belarusian State University, 4 Nezavisimosti Avenue, Minsk 200030, Belarus
e-mail: bosiakov@bsu.by; mikhasev@bsu.by; rogosin@bsu.by; rogosinsv@gmail.com

P. Drygaś, S. Rogosin (eds.), *Modern Problems in Applied Analysis*,
Trends in Mathematics, https://doi.org/10.1007/978-3-319-72640-3_4

1 Introduction

A tooth root is attached to alveolar bone by a periodontal ligament (PDL), a soft connective tissue consisting of collagen fibres and a matrix phase with nerve endings and blood vessels [1, 2]. Because of the low stiffness, the PDL plays a central role in the tooth mobility [3] and acts basically as a physiological mechanism responsible for the teeth movements and teeth reaction to loading [4–6]. The long-term force action on the tooth is characterized by its nearly instantaneous displacement in the PDL with a subsequent relaxation of the stresses during five hours of the load action [7], showing that the PDL reveals the viscoelastic and time-dependent properties [8–13].

An analytical model of the initial (instantaneous) translational displacements of the tooth root in the PDL under the action of static load has been proposed in [14]. This model has several advantages compared with other known analytical models [15, 16]. These advantages are the possibility to use arbitrary elastic constants for the PDL tissue and assess stresses in the PDL without simplifying assumptions. The static model developed in [14] permits also to evaluate the stress-strain state of the PDL throughout its thickness. The direct and logical evolution of this model seems to be taking into account the viscoelastic properties of the periodontal ligament. Such a model allowing for both the stress relaxation and slow displacement of teeth could be adopted in the real clinical practice for the bone reconstruction during the orthodontic treatment [17] and for elastic composites [18].

Analytical modelling of the PDL viscoelastic response under the tooth long-term loading were carried out in [19–21]. In particular, in [19], a nonlinear viscoelastic constitutive model was adopted to describe the relaxation phenomena in the PDL which is consistent with experimental data describing the dependence of relaxation rate on the level of an applied strain. The proposed model allows one to evaluate with sufficient accuracy the PDL behavior under the high rate loading and large strains of the PDL tissue. At the same time, modelling of the tooth roots motions in the viscoelastic PDL was not carried out in [19]. The mathematical model with single degree of freedom for defining the initial posterior tooth displacement in the viscoelastic PDL associated with functioning interproximal contacts was developed in paper [20]. This model is described by ordinary differential equations with constant coefficients and allows one to simulate the short-term movements of the teeth at the occurrence of the contact between the adjacent teeth. In [21] a mathematical model for description of experimentally observed viscoelastic and time-dependent behaviours of the PDL was developed. The analysis is focused on the evolution of translational displacements of the tooth root in the PDL under the vertical load (intrusion). The calculated tooth-root displacement with time at a constant load allowed comparing the behaviour of the viscoelastic model with the fractional exponential kernel with that of the known nonlinear viscoelastic model of the tooth-root movement developed in [12, 20]. The model proposed in [21] generalizes the known analytical models of the viscoelastic PDL by introduction the instantaneous and relaxed elastic moduli, as well as the fractional parameter.

The advantage of this model is in the use of the fractional parameter improving the description of various pathological processes and age-related changes in the PDL. The fractional parameter makes it possible to take into account different behaviours of the periodontal tissue under short- and long-term loads. For instance, it allows assessing the change in the time interval of a transition phase for a given maximum displacement.

The aim of this study is to develop a 2-D analytical model of the PDL viscoelastic response which would permit to predict the stress-strain state of the PDL characterized by the phenomena of relaxation and the continuing translational movement of the tooth root as well. A 2-D model is quite sufficient to describe the PDL behavior during the translational movement of the tooth root, with the exception of the small PDL regions near the root apex and the alveolar crest. This agrees with the conclusions presented in [15, 16], showing that the vertical displacements of the PDL points are very small and can be neglected, with the exception displacements of the PDL points at the regions near the apex and the alveolar crest.

2 Viscoelastic Model of the PDL

It is assumed that at the initial moment the tooth root is shifted horizontally at the distance u_0. The cross-section of the tooth root is circular in the any plane perpendicular to the longitudinal axis of the tooth. The tooth root is considered as the rigid body compared with the PDL tissue [22]. Positions of the root section before and after the displacement, as well as the geometric dimensions of the root cross-section are shown in Fig. 1.

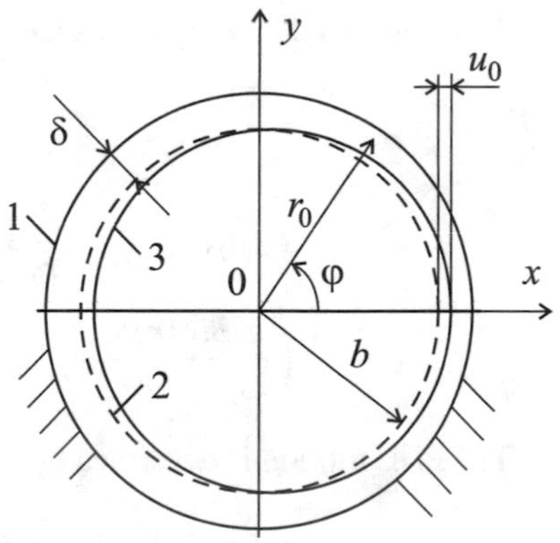

Fig. 1 Positions of tooth root in sectional plane before and after displacement: 1—outer contour of the PDL fixed on dental alveolus surface; 2—position of cross-section of tooth root before load action; 3—position of cross-section of tooth root after load action; u_0 is the tooth root displacement in the x-direction; r_0 is the radial coordinate, φ is the polar angle

In the polar coordinate system r_0, φ (see Fig. 1), the constitutive equations for the viscoelastic PDL are taken in the form:

$$\sigma_{rr}(r_0,\varphi,t) = \lambda\left(e_{\varphi\varphi}(r_0,\varphi,t) - \int_0^t K(t-\tau)e_{\varphi\varphi}(r_0,\varphi,\tau)d\tau\right) +$$
$$(\lambda+2\mu)\left(e_{rr}(r_0,\varphi,t) - \int_0^t K(t-\tau)e_{rr}(r_0,\varphi,\tau)d\tau\right),$$
$$\sigma_{r\varphi}(r_0,\varphi,t) = 2\mu\left(e_{r\varphi}(r_0,\varphi,t) - \int_0^t K(t-\tau)e_{r\varphi}(r_0,\varphi,\tau)d\tau\right), \tag{2.1}$$
$$\sigma_{\varphi\varphi}(r,\varphi,t) = \lambda\left(e_{rr}(r_0,\varphi,t) - \int_0^t K(t-\tau)e_{rr}(r,\varphi,\tau)d\tau\right) +$$
$$(\lambda+2\mu)\left(e_{\varphi\varphi}(r_0,\varphi,t) - \int_0^t K(t-\tau)e_{\varphi\varphi}(r_0,\varphi,\tau)d\tau\right),$$

where $\sigma_{rr}, \sigma_{r\varphi}, \sigma_{\varphi\varphi}$ and $e_{rr}, e_{r\varphi}, e_{\varphi\varphi}$ are the components of the stress and strain tensors, respectively; $\lambda = \frac{E_0\nu}{(1-2\nu)(1+\nu)}$ and $\mu = \frac{E_0}{2(1+\nu)}$ are Lame parameters; E_0 and ν are the instantaneous modulus elasticity and Poisson's ratio of the PDL (ν =const); $K(t)$ is the relaxation kernel for the normal and shear components of stresses; $u_r(r_0,\varphi,t)$, $u_\varphi(r_0,\varphi,t)$ are the radial and circular displacements of the PDL points, respectively.

The strain tensor components are

$$e_{rr}(r_0,\varphi,t) = \frac{\partial u_r(r_0,\varphi,t)}{\partial r},$$
$$e_{\varphi\varphi}(r_0,\varphi,t) = \frac{1}{r_0}\left(\frac{\partial u_\varphi(r_0,\varphi,t)}{\partial\varphi} + u_r(r_0,\varphi,t)\right), \tag{2.2}$$
$$e_{r\varphi} = \frac{1}{2}\left(\frac{1}{r_0}\frac{\partial u_r(r_0,\varphi,t)}{\partial\varphi} + \frac{\partial u_\varphi(r_0,\varphi,t)}{\partial r} - \frac{u_\varphi(r_0,\varphi,t)}{r_0}\right).$$

The PDL motion is described by the following equations:

$$\frac{\partial\sigma_{rr}}{\partial r_0} + \frac{1}{r_0}\frac{\partial\sigma_{r\varphi}}{\partial\varphi} + \frac{\sigma_{rr}-\sigma_{\varphi\varphi}}{r_0} = \rho\frac{\partial^2 u_r(r_0,\varphi,t)}{\partial t^2},$$
$$\frac{1}{r_0}\frac{\partial\sigma_{\varphi\varphi}}{\partial\varphi} + \frac{\partial\sigma_{r\varphi}}{\partial r_0} + \frac{2\sigma_{r\varphi}}{r_0} = \rho\frac{\partial^2 u_\varphi(r_0,\varphi,t)}{\partial t^2}, \tag{2.3}$$

where ρ is the density of the PDL tissue.

Substituting (2.1) and (2.2) into Eqs. (2.3) leads to the following system of integro-differential equations:

$$\left(r_0\frac{\partial^2}{\partial r_0^2}+\frac{\partial}{\partial r_0}-\frac{1}{r_0}+\frac{\mu}{\lambda+2\mu}\frac{1}{r_0}\frac{\partial^2}{\partial \varphi^2}\right)$$

$$\left(u_r(r_0,\varphi,t)-\int_0^t K(t-\tau)u_r(r_0,\varphi,\tau)d\tau\right)+$$

$$+\left(\frac{\lambda+\mu}{\lambda+2\mu}\frac{\partial^2}{\partial\varphi\partial r_0}-\frac{\lambda+3\mu}{\lambda+2\mu}\frac{1}{r_0}\frac{\partial}{\partial\varphi}\right)$$

$$\left(u_\varphi(r_0,\varphi,t)-\int_0^t K(t-\tau)u_\varphi(r_0,\varphi,\tau)d\tau\right)=\rho\frac{\partial^2 u_r(r_0,\varphi,t)}{\partial t^2},$$

$$\left(\frac{\lambda+\mu}{\lambda+2\mu}\frac{\partial^2}{\partial\varphi\partial r_0}+\frac{\lambda+3\mu}{\lambda+2\mu}\frac{1}{r_0}\frac{\partial}{\partial\varphi}\right)$$

$$\left(u_r(r_0,\varphi,t)-\int_0^t K(t-\tau)u_r(r_0,\varphi,\tau)d\tau\right)+$$

$$+\left(\frac{\mu r_0}{\lambda+2\mu}\frac{\partial^2}{\partial r_0^2}+\frac{\mu}{\lambda+2\mu}\frac{\partial}{\partial r_0}-\frac{\mu}{\lambda+2\mu}\frac{1}{r_0}+\frac{1}{r_0}\frac{\partial^2}{\partial\varphi^2}\right)$$

$$\left(u_\varphi(r_0,\varphi,t)-\int_0^t K(t-\tau)u_\varphi(r_0,\varphi,\tau)d\tau\right)=\rho\frac{\partial^2 u_\varphi(r_0,\varphi,t)}{\partial t^2}.$$

The above equations can be rewritten in the dimensionless form:

$$\begin{gathered}
a_{11}\left(u(r,\varphi,t)-\int_0^t K(t-\tau)u(r,\varphi,\tau)d\tau\right)+\\
a_{12}\left(v(r,\varphi,t)-\int_0^t K(t-\tau)v(r,\varphi,\tau)d\tau\right)=\varepsilon\frac{\partial^2 u(r,\varphi,t)}{\partial t^2},\\
a_{21}\left(u(r,\varphi,t)-\int_0^t K(t-\tau)u(r,\varphi,\tau)d\tau\right)+\\
a_{22}\left(v(r,\varphi,t)-\int_0^t K(t-\tau)v(r,\varphi,\tau)d\tau\right)=\varepsilon\frac{\partial^2 v(r,\varphi,t)}{\partial t^2},
\end{gathered}\tag{2.4}$$

where

$$a_{11} = r(1-\nu)\frac{\partial^2}{\partial r^2} + (1-\nu)\frac{\partial}{\partial r} - \frac{1-\nu}{r} + \frac{1-2\nu}{2}\frac{1}{r}\frac{\partial^2}{\partial \varphi^2},$$

$$a_{12} = \frac{1}{2}\frac{\partial^2}{\partial \varphi \partial r} - \frac{3-4\nu}{2}\frac{1}{r}\frac{\partial}{\partial \varphi}, a_{21} = \frac{1}{2}\frac{\partial^2}{\partial \varphi \partial r} + \frac{3-4\nu}{2}\frac{1}{r}\frac{\partial}{\partial \varphi},$$

$$a_{22} = \frac{(1-2\nu)r}{2}\frac{\partial^2}{\partial r^2} + \frac{1-2\nu}{2}\frac{\partial}{\partial r} - \frac{1-2\nu}{r} + \frac{1-\nu}{r}\frac{\partial^2}{\partial \varphi^2}.$$

Here $u(r,\varphi,t) = \frac{u_r(r,\varphi,t)}{h}$, $v(r,\varphi,t) = \frac{u_\varphi(r,\varphi,t)}{h}$ are the dimensionless displacements, $r = \frac{r_0}{h}$ is the dimensionless radial coordinate, h is the tooth root height, $\varepsilon = \frac{\rho h^2}{E_0 t_0^2}(1+\nu)(1-2\nu)$ is a dimensionless parameter, and t_0 is the characteristic time.

Let us observe the values of parameters in our model. Overview of the ranges of Poisson's ratios and elastic moduli for the PDL indicates that their lowest values are 0.28 and 0.01 MPa, respectively [1, 23–25]. The density of the PDL tissue is equal to 1.06 g/cm^3 [29]. The tooth root height is assumed to be 13.0 mm, the thickness of the PDL along the normal to the inner surface is about 0.229 mm [15]. Commonly used the elastic constants are $E_0 = 680$ kPa and $\nu = 0.49$ [15, 16, 26, 27]. For the geometric and time-material constants mentioned above, the small parameter ε amounts approximately to 6.06×10^{-16}.

It is assumed that during the tooth root translational movement in the PDL, the required displacements can be represented as products of independent functions

$$u(r,\varphi,t) = u_1(r,t)\cos(\varphi), \quad v(r,\varphi,t) = v_1(r,t)\sin(\varphi). \tag{2.5}$$

Inserting (2.5) into Eq. (2.4) yields the following system of equations (which do not depend on the polar angle due to the form of the operator coefficients):

$$\begin{gathered}
A_{11}\left(u_1(r,t) - \int_0^t K(t-\tau)u_1(r,\tau)d\tau\right) + \\
A_{12}\left(v_1(r,t) - \int_0^t K(t-\tau)v_1(r,\tau)d\tau\right) = \varepsilon\frac{\partial^2 u_1(r,t)}{\partial t^2}, \\
A_{21}\left(u_1(r,t) - \int_0^t K(t-\tau)u_1(r,\tau)d\tau\right) + \\
A_{22}\left(v_1(r,t) - \int_0^t K(t-\tau)v_1(r,\tau)d\tau\right) = \varepsilon\frac{\partial^2 v_1(r,t)}{\partial t^2}.
\end{gathered} \tag{2.6}$$

with the differential operators A_{ij} defined as

$$A_{11} = r(1-\nu)\frac{\partial^2}{\partial r^2} + (1-\nu)\frac{\partial}{\partial r} - \frac{3-4\nu}{2r},$$
$$A_{12} = \frac{1}{2}\frac{\partial}{\partial r} - \frac{3-4\nu}{2}\frac{1}{r}, A_{21} = -\frac{1}{2}\frac{\partial}{\partial r} - \frac{3-4\nu}{2}\frac{1}{r},$$
$$A_{22} = \frac{(1-2\nu)r}{2}\frac{\partial^2}{\partial r^2} + \frac{1-2\nu}{2}\frac{\partial}{\partial r} - \frac{3-4\nu}{2r}.$$

Let us apply the Laplace transform to Eq. (2.6) with the initial conditions

$$\begin{gathered} u_1(r,0) = u_2(r),\ \ v_1(r,0) = v_2(r), \\ \left.\frac{\partial u_1(r,t)}{\partial t}\right|_{t=0} = 0,\ \ \left.\frac{\partial v_1(r,t)}{\partial t}\right|_{t=0} = 0, \end{gathered} \tag{2.7}$$

where $u_2(r)$ and $v_2(r)$ are the initial displacements of the PDL points emerging after the instantaneous shift of the tooth root by the distance u_0 (see paper [14]).

As a result, one obtains the system of differential equations

$$\begin{gathered} B_{10}\left(B_{11} + B_{12}\frac{\partial}{\partial r} + B_{13}\frac{\partial^2}{\partial r^2}\right)u_1^\star(r,p) + \\ \left(B_{14} + B_{15}\frac{\partial}{\partial r}\right)v_1^\star(r,p) = 0, \\ B_{20}\left(B_{21} + B_{22}\frac{\partial}{\partial r} + B_{23}\frac{\partial^2}{\partial r^2}\right)v_1^\star(r,p) + \\ \left(B_{24} + B_{25}\frac{\partial}{\partial r}\right)u_1^\star(r,p) = 0, \end{gathered} \tag{2.8}$$

with respect to the functions $u_1^\star(r,p)$, $v_1^\star(r,p)$ being the Laplace transforms of $u_1(r,t)$, $v_1(r,t)$, respectively. Here

$$B_{10} = -p\varepsilon u_1(r), B_{11} = \varepsilon p^2 - \frac{f(p)(3-4\nu)}{2r^2},$$
$$B_{12} = \frac{f(p)(1-\nu)}{r}, B_{13} = f(p)(1-\nu), B_{14} = -\frac{f(p)(3-4\nu)}{2r^2},$$
$$B_{15} = \frac{f(p)}{2r}, B_{20} = -p\varepsilon v_1(r), B_{21} = \varepsilon p^2 - \frac{f(p)(3-4\nu)}{2r^2},$$
$$B_{22} = \frac{f(p)(1-2\nu)}{2r}, B_{23} = \frac{f(p)(1-2\nu)}{2}, B_{24} = -\frac{f(p)(3-4\nu)}{2r^2},$$
$$B_{25} = -\frac{f(p)}{2r}, f(p) = -1 + K^\star(p),\quad K^\star = \int_0^\infty K(t)\exp(-pt)\mathrm{d}t.$$

It should be noted that Eq. (2.8) turn out to be regularly perturbed by a small parameter ε. Their solution could be sought in the form of asymptotic series:

$$u_1^\star(r,p) = \sum_{k=1}^{\infty} u_{0k}^\star(r,p)\varepsilon^{k-1}, \, v_1^\star(r,p) = \sum_{k=1}^{\infty} v_{0k}^\star(r,p)\varepsilon^{k-1}. \tag{2.9}$$

The substitution of expansions (2.9) into Eq. (2.8) gives us a sequence of differential equations for functions $u_{0k}^\star$, $v_{0k}^\star$. Let us consider here only the first two approximations. They produce the following two systems:

$$\begin{aligned} C_{11}u_{01}^\star(r,p) + C_{12}v_{01}^\star(r,p) &= 0, \\ C_{21}u_{01}^\star(r,p) + C_{22}v_{01}^\star(r,p) &= 0, \end{aligned} \tag{2.10}$$

and

$$\begin{aligned} -p(u_2(r) - pu_{01}^\star(r,p)) + C_{11}u_{02}^\star(r,p) + C_{12}v_{02}^\star(r,p) &= 0, \\ -p(v_2(r) - pv_{01}^\star(r,p)) + C_{21}u_{02}^\star(r,p) + C_{22}v_{02}^\star(r,p) &= 0, \end{aligned} \tag{2.11}$$

where

$$C_{11} = -\frac{3-4\nu}{2r^2} + \frac{1-\nu}{r}\frac{\partial}{\partial r} + (1-\nu)\frac{\partial^2}{\partial r^2}, C_{12} = \frac{1}{2r}\left(-\frac{3-4\nu}{r} + \frac{\partial}{\partial r}\right),$$

$$C_{21} = \frac{1}{2r}\left(-\frac{3-4\nu}{r} - \frac{\partial}{\partial r}\right), C_{22} = -\frac{3-4\nu}{2r^2} + \frac{1-2\nu}{2r}\frac{\partial}{\partial r} + \frac{1-2\nu}{2}\frac{\partial^2}{\partial r^2}$$

are the differential operators.

The solution to Eq. (2.10) is the following [14]:

$$\begin{aligned} u_{01}^\star(r,p) &= c_{01}(1-4\nu)r^2 + \frac{c_{02}}{r} - \frac{c_{03}}{2(3-4\nu)} + c_{03}\ln(r) + c_{04}, \\ v_{01}^\star(r,p) &= c_{01}(5-4\nu)r^2 + \frac{c_{02}}{r} - \frac{c_{03}}{2(3-4\nu)} - c_{03}\ln(r) - c_{04}, \end{aligned} \tag{2.12}$$

where c_{0k} are unknown constants, $k = \overline{1,4}$. Taking the initial displacements $u_2(r)$ and $v_2(r)$, into account the solution of Eq. (2.11) can be represented in the form of functions

$$\begin{aligned} u_{02}^\star(r,p) &= b_{13}r^4 + b_{23}r^2\ln(r) + b_{33} + b_{43}r^2, \\ v_{02}^\star(r,p) &= f_{13}r^4 + f_{23}r^2\ln(r) + f_{33} + f_{43}r^2 \end{aligned} \tag{2.13}$$

with the coefficients

$$b_{13} = -\frac{p(pc_{01} - p_{01})(1 - 12\nu + 16\nu^2)}{24(1 - 3\nu + 2\nu^2)f^\star(p)}, f_{13} = -\frac{p(pc_{01} - p_{01})(19 - 36\nu + 16\nu^2)}{24(1 - 3\nu + 2\nu^2)f^\star(p)},$$

$$b_{33} = f_{33} = \frac{p(pc_{02} - p_{02})}{(3 - 4\nu)f^\star(p)}, b_{23} = -\frac{p(c_{03} - p_{03})(7 - 20\nu + 16\nu^2)}{8(3 - 13\nu + 18\nu^2 - 8\nu^3)f^\star(p)},$$

$$f_{23} = \frac{2p(pc_{03} - p_{03}) - b_{23}(21 - 52\nu + 32\nu^2)f^\star(p)}{(15 - 44\nu + 32\nu^2)f^\star(p)},$$

$$b_{43} = f_{43} = \frac{p\left((pc_{03} - p_{03})(17 - 48\nu + 32\nu^2) - 16(pc_{04} - p_{04})(1 - 3\nu + 2\nu^2)\right)}{32(1 - 3\nu + 2\nu^2)f^\star(p)}.$$

Unknown constants c_{0k}, $k = \overline{1, 4}$ are determined numerically via the inverse Laplace transform of the solutions of Eq. (2.9) and functions (2.12), (2.13) subjected to the following boundary conditions:

$$u_1(b, 0) = u_0, \quad v_1(b, 0) = u_0, \quad u_1(b_1, 0) = 0, \quad v_1(b, 0) = 0. \tag{2.14}$$

The inverse Laplace transformation was performed using the geometric dimensions b and b_1 of the cross-section corresponding to the tooth root in the shape of a circular paraboloid; the radius of the inner tooth surface at the alveolar crest level was 3.9 mm [15]. The viscoelastic properties of the PDL are described by model with fractional derivatives [30] using the fractional exponential function introduced by Rabotnov [32, 33]:

$$\mathcal{E}_\gamma\left(-\frac{t}{\tau_\varepsilon}\right) = \frac{t^{\gamma-1}}{\tau_\varepsilon^\gamma}\sum_{n=0}^{\infty}(-1)^n\frac{(t/\tau_\varepsilon)^{\gamma n}}{\Gamma[\gamma(n+1)]}, \tag{2.15}$$

where $0 < \gamma < 1$ is the fractional parameter. In accordance with study [7], it was suggested that the stress relaxation in the PDL after starting the load action (and subsequent instantaneous displacement of the tooth root in the PDL) occurred within about five hours.

2.1 *Evolution of Stresses in the PDL*

Elastic constants for the PDL were taken to be equal to $E = 680$ kPa and $\nu = 0.49$ [15, 16, 26, 27]; the height of the tooth root was 13.0 mm, thickness of the PDL along normal to its inner surface was 0.229 mm [15]. The radii of the inner and outer surface of the PDL corresponded to the middle third of the tooth root in the form of a circular paraboloid (radius of the tooth cross-section at the alveolar crest level was equal to 3.9 mm). The load applied to the tooth was 1 N; due to this load, the instantaneous dimensionless initial displacement was $u_0 = 18.15 \times 10^{-6}$ [14]. The parameter γ and retardation time τ_ε for the model of viscoelasticity model with fractional exponential kernel were taken as 0.3 and 550 s, respectively [21].

Patterns of the radial normal component $\sigma_{rr}(r, \varphi, t_k)$ of stresses in the PDL for points of time $t_1 = 0.25$ h, $t_2 = 1$ h, $t_3 = 2$ h and $t_4 = 5$ h for these two cases are shown in Fig. 2 ($-\pi/2 \le \varphi \le \pi/2$).

It is seen from Fig. 2 that the radial stresses throughout the PDL thickness change insignificantly. For any point of time and for the fixed radial coordinate, the circular stresses at the PDL inner surface turn out to be by 0.7% more than the radial stresses at the PDL outer surface. This outcome suggests that the analytical model of the nearly incompressible PDL (with Poisson's ratio of about 0.49) developed in studies [15, 16] can be used with high accuracy to estimate stresses throughout the PDL thickness. The ratio between the respective stresses at the inner and outer surface of the PDL is retained when the load changes. The stresses pattern (see Fig. 2) corresponds to the compressive region ($\pi/2 \le \varphi \le 3\pi/2$) of the PDL located in the direction of the tooth root movement. The patterns of the tensile radial stresses with respect to the coordinate origin are symmetric to the corresponding patterns of the compressive stresses for the same point of time. The maximum and minimum radial stresses occure on the PDL inner surface at $\varphi = \pi$ and $\varphi = 0$, respectively. The patterns of the circular component $\sigma_{\varphi\varphi}$ of stresses on the radial and circular coordinates for various points of time are similar to the patterns shown in Fig. 2.

Normal and tangential stresses in the PDL (material constants and geometric parameters of the PDL are the same) increase insignificantly with increasing of the fractional parameter; the highest stresses correspond to the fractional parameter $\gamma = 1$. The patterns of radial stresses in the PDL for different fractional parameters are similar to the patterns shown in Fig. 2.

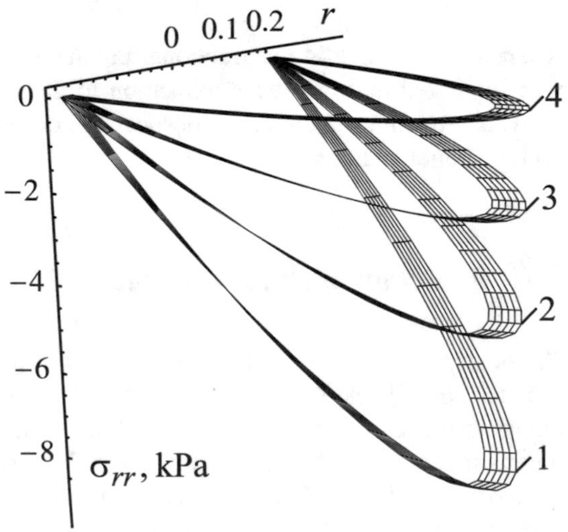

Fig. 2 Patterns of the radial normal components σ_{rr} of stresses in the PDL for different point of time: 1—$t = 0.25$ h; 2—$t = 1$ h; 3—$t = 2$ h; 4—$t = 5$ h

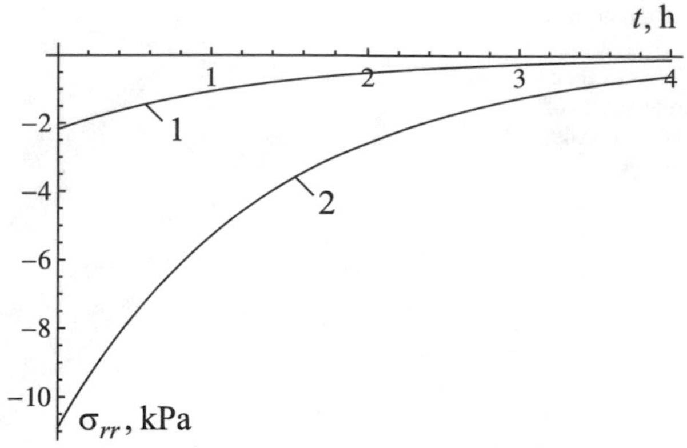

Fig. 3 Radial component $\sigma_{rr}(b, 0, t)$ of stresses vs time for the point on the PDL inner surface at $\varphi = 0$: 1—$u_0 = 18.15 \times 10^{-6}$ (correspond to load of 1 N applied to crown of the tooth root); 2—$u_0 = 90.75 \times 10^{-6}$ (correspond to load of 5 N applied to crown of the tooth root)

Figure 3 shows the radial component σ_{rr} of stresses versus time for a point on the PDL inner surface at $\varphi = 0$ for different initial displacements. The geometric and material constants as well as the conditions of the load application were the same.

It can be seen from Fig. 3 the larger the applied load, the longer period of time for the total relaxation of stresses in the PDL is required.

2.2 *Effect of Poisson's Ratio on the Normal Component of Stresses*

Frequently used Poisson's ratios, along with 0.49, are taken to be equal 0.30 or 0.45 [1, 23–25]. Adopting different Poisson's ratios from such a large range can significantly affect the results of the finite-element and analytical modelling of the PDL behavior under the static and dynamic loading. So, results presented in [14] have revealed that the instantaneous initial displacement u_0 could change considerably (about to four times) at the change of Poisson's ratio from 0.30 to 0.49 (cf. [27, 28]). It has been also found that the Poisson's ratio significantly affects stress components, as well as the stress regimes in the linearly elastic PDL. At the same time, the influence of the Poisson's ratio variation on the time-dependent stresses in the viscoelastic PDL was not analyzed. The patterns of the normal radial stresses in the PDL with Poisson's ratio 0.30, 0.45 and 0.49 at the point of time 2 h are shown in Fig. 4. The geometric and other material constants as well as the load were taken as above.

Figure 4 shows that the Poisson's ratio significantly affects both the magnitude of the radial stresses in the PDL and the stress distributions. The decrease in the magnitude of Poisson's ratio results in the decrease in the magnitude of the stress.

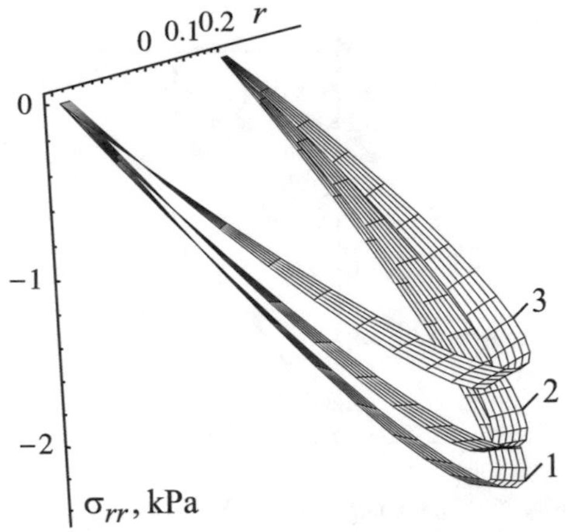

Fig. 4 Patterns of the normal radial stresses $\sigma_{rr}(r, \varphi, t_3)$ in the PDL for the point of time $t_3 = 2$ h at different values of Poisson's ratio: 1—$\nu = 0.49$; 2—$\nu = 0.45$; 3—$\nu = 0.30$

At the same time, decrease in Poisson's ratio leads to the increase in the difference between the radial stresses at the inner and outer PDL surfaces. So, for Poisson's ratio $\nu = 0.49$, this difference is 0.7%; for Poisson's ratios $\nu = 0.45$ and $\nu = 0.30$, they are about 3.1% and 9.5%, respectively.

3 Conclusions

In this paper, the analytical model of the viscoelastic PDL based on fractional viscoelastic model was proposed. It allows one to assess the time-dependent stresses in the PDL after the tooth root translational displacement. Parameters for the kernel of relaxation were defined under the assumption that the relaxation of stresses in the PDL tissue occurs in about 5 h after definition of the initial displacement of the tooth root. The analysis of performed calculations allows us to make the following conclusions:

- The normal radial and circular components of stresses decrease throughout the thickness of the PDL (from its inner surface to the outer one). Damping of stresses occurs faster for small initial displacement; the larger initial displacement is, the more time for the stress relaxation is required.
- If the PDL is nearly incompressible (with Poisson's ratio $\nu = 0.49$), the normal stresses are changed significantly over the PDL width. The value of Poisson's ratio significantly affects stresses in the PDL tissues. The normal stresses decrease together with Poisson's ratio in so doing, the difference between stresses on the inner and outer surfaces of the PDL increases. Therefore, Poisson ratio may significantly effect on the outcomes of the practical assessment of the PDL behavior based on the linear elastic and viscoelastic models.

- Further development of the proposed model can be associated with the 3-D analytical modeling which would enable one to describe the dynamic behavior of the viscoelastic PDL under various loads applied to the tooth root taking its real 3-D shape into account.

The outcomes obtained in this study can be used for simulation of the bone remodelling process during the orthodontic treatment and for assessment of the optimal conditions for the orthodontic load application.

Acknowledgements This paper is the result of project implementation: TAMER "Trans-Atlantic Micromechanics Evolving Research: Materials containing inhomogeneities of diverse physical properties, shapes and orientations" supported by FP7-PEOPLE-2013-IRSES Marie Curie Action "International Research Staff Exchange Scheme". It is also supported by Belarusian Fund for Fundamental Scientific Research (grant F17MS-002).

References

1. T.S. Fill, J.P. Carey, R.W. Toogood, P.W. Major, Experimentally determined mechanical properties of, and models for, the periodontal ligament: critical review of current literature. J. Dent. Biomech. (2011). https://doi.org/10.4061/2011/312980
2. A.N. Natali, A.R. Ten Cate, *Dental Biomechanics* (Taylor and Francis, London, 2003)
3. C. Dorow, F.G. Sander, Development of a model for the simulation of orthodontic load on lower first premolars using the finite element method. J. Orofac. Orthop. **66**, 208–218 (2005)
4. M. Drolshagen, L. Keilig, I. Hasan, S. Reimann, J. Deschner, K.T. Brinkmann, R. Krause, M. Favino, C. Bourauel, Development of a novel intraoral measurement device to determine the biomechanical characteristics of the human periodontal ligament. J. Biomech. **44**, 2136–2143 (2011)
5. F. Groning, M.J. Fagan, P. O'Higins, The effects of the periodontal ligament on mandibular stiffness: a study combining finite element analysis and geometric morphometrics. J. Biomech. **44**, 1304–1312 (2011)
6. W.R. Proffit, H.W. Fields, D.M. Sarver, *Contemporary Orthodontics* (Mosby, St. Louis, 2012)
7. W.D. Van Driel, E.J. Van Leewen, J.W. Von den Hoff, J.C. Maltha, A.M. Kuijpers-Jagtman, Time-dependent mechanical behavior of the periodontal ligament. Proc. Inst. Mech. Eng. Part H J. Eng. Med. **214**, 407–504 (2000)
8. C.J. Burstone, R.J. Pryputniewicz, W.W. Bowley, Holographic measurement of tooth mobility in three dimensions. J. Periodontal Res. **13**, 283–294 (1978)
9. K. Komatsu, Mechanical strength and viscoelastic response of the periodontal ligament in relation to structure. J. Dent. Biomech. (2010). https://doi.org/10.4061/2010/502318
10. J. Middleton, M. Jones, A. Wilson, The role of the periodontal ligament in bone modeling: the initial development of a time-dependent finite element model. Am. J. Orthod. Dentofac. Orthop. **109**, 155–162 (1996)
11. D.C.A. Picton, Tooth mobility – an update. Eur. J. Orthod. **12**, 109–115 (1990)
12. L. Qian, M. Todo, Y. Morita, Y. Matsushita, K. Koyano, Deformation analysis of the periodontium considering the viscoelasticity of the periodontal ligament. Dent. Mater. **25**, 1285–1292 (2009)
13. S.R. Toms, G.J. Dakin, J.E. Lemons, A.W. Eberhardt, Quasi-linear viscoelastic behavior of the human periodontal ligament. J. Biomech. **35**, 1411–1415 (2002)
14. S. Bosiakov, G. Mikhasev, Mathematical model for analysis of translational displacements of tooth root. Math. Model. Anal. **20**, 490–501 (2015)
15. C.G. Provatidis, An analytical model for stress analysis of a tooth in translation. Int. J. Eng. Sci. **39**, 1361–1381 (2001)

16. A. Van Schepdael, L. Geris, J. Van der Sloten, Analytical determination of stress patterns in the periodontal ligament during orthodontic tooth movement. Med. Eng. Phys. **35**, 403–410 (2013)
17. V. Uchaikin, *Fractional Derivatives for Physicists and Engineers*. Applications, vol. II (Springer/Higher Education Press, Berlin/Beijing, 2013)
18. P. Drygaś, Functional-differential equations in a class of analytic functions and its application to elastic composites. Compl. Var. Ellipt. Equat. **61**, 1145–1156 (2016)
19. A.N. Natali, P.G. Pavan, C. Venturato, K. Komatsu, Constitutive modeling of the non-linear visco-elasticity of the periodontal ligament. Comput. Methods Prog. Biomed. **104**, 193–198 (2011)
20. N. Slomka, A.D. Vardimon, A. Gefen, R. Pilo, C. Bourauel, T. Brosh, Time-related PDL: viscoelastic response during initial orthodontic tooth movement of a tooth with functioning interproximal contact – a mathematical model. J. Biomech. **41**, 1871–1877 (2008)
21. S. Bosiakov, A. Koroleva, S. Rogosin, V. Silberschmidt, Viscoelasticity of periodontal ligament: an analytical model. Mech. Adv. Mater. Mod. Proc. **1**(7) (2015). https://doi.org/10.1186/s40759-015-0007-0
22. A. Hohmann, C. Kober, P. Young, C. Dorow, M. Geiger, A. Boryor, F.M. Sander, C. Sander, F.G. Sander, Influence of different modeling strategies for the periodontal ligament on finite element simulation results. Am. J. Orthod. Dentofac. Orthop. **139**, 775–783 (2011)
23. A. Kawarizadeh, C. Bourauel, A. Jager, Experimental and numerical determination of initial tooth mobility and material properties of the periodontal ligament in rat molar specimens. Eur. J. Orthod. **25**, 569–578 (2003)
24. M. Poppe, C. Bourauel, A. Jager, Determination of the material properties of the human periodontal ligament and the position of the centers of resistance in single-rooted teeth. J. Orofac. Orthop. **64**, 358–370 (2002)
25. J.S. Rees, P.H. Jacobsen, Elastic modulus of the periodontal ligament. Biomaterials **18**, 995–999 (1997)
26. K. Tanne, T. Nagataki, Y. Inoue, M. Sakuda, C.J. Burstone, Patterns of initial tooth displacements associated with various root lengths and alveolar bone heights. Am. J. Orthod. Dentofac. Orthop. **100**, 66–71 (1991)
27. K. Tanne, M. Sakuda, C.J. Burstone, Three-dimensional finite element analysis for stress in the periodontal tissue by orthodontic forces. Am. J. Orthod. Dentofac. Orthop. **92**, 499–505 (1987)
28. R.F. Viecilli, Th.R. Katona, J. Chen, J.K. (Jr.) Hartsfield, W.E. Roberts, Three-dimensional mechanical environment of orthodontic tooth movement and root resorption. Am. J. Orthod. Dentofac. Orthop. **133**, 791.e11–791.e26 (2008)
29. S.A. Wood, D.C. Strait, E.R. Dumont, C.F. Ross, I.R. Grosse, The effects of modeling simplifications on craniofacial finite element models: the alveoli (tooth sockets) and periodontal ligaments. J. Biomech. **44**, 1831–1838 (2011)
30. Yu.A. Rossikhin, M.V. Shitikova, Comparative analysis of viscoelastic models involving fractional derivatives of different orders. Frac. Calc. Appl. Anal. **10**, 111–121 (2007)
31. Yu. A. Rossikhin, M. V. Shitikova, Two approaches for studying the impact response of viscoelastic engineering systems: an overview. Comp. Math. Appl. **66**, 755–773 (2013)
32. Yu.N. Rabotnov, Equilibrium of an elastic medium with after-effect. J. Appl. Math. Mech. **12**, 53–62 (1948)
33. Yu.N. Rabotnov, *Elements of Hereditary Solid Mechanics* (Mir Publishers, Moscow, 1980)

Dirichlet Type Problems in Polydomains

A. Okay Çelebi

Abstract In this paper, we investigate a Dirichlet type problem, known as Riquier problem, for higher order linear complex differential equations in the unit polydisc of $\mathbb{C}^2$. After deriving a Green's function, we present the solution for a model equation with homogeneous boundary conditions. Afterwards we obtain the solution of a linear equation for Riquier boundary value problem on the unit polydisc in $\mathbb{C}^2$.

Keywords Dirichlet type problem • Riquier problem • Polydisc • Complex partial differential equations

Mathematics Subject Classification (2010) Primary 32W50; Secondary 32A26

1 Introduction

Dirichlet type problems in $\mathbb{C}$ with polyharmonic or polyanalytic principle parts are investigated in a variety of domains by many researchers in the unit disc of the complex plane [2, 4–7, 11–13, 22], in the half plane [9], circular rings [10, 21] and for inhomogeneous linear complex equations with the polyharmonic operators as leading terms [1]. The partial differential equations in several complex variables have attracted researchers in the last several decades. For an introduction to various types of boundary value problems we refer, for example, the book by Begehr-Dzhuraev [8]. Higher-order integral representations in polydomains have been considered by Krauss [14, 15]. In the recent years Riemann, Riemann-Hilbert, Schwarz, Neumann and Robin problems are treated in relation with pluriholomorphic and pluriharmonic functions [16, 18, 19]. Recently we have considered the Schwarz problem in a polydisc in $\mathbb{C}^2$ [3].

A.O. Çelebi (✉)
Department of Mathematics, Yeditepe University, Istanbul, Turkey
e-mail: acelebi@yeditepe.edu.tr

P. Drygaś, S. Rogosin (eds.), *Modern Problems in Applied Analysis*,
Trends in Mathematics, https://doi.org/10.1007/978-3-319-72640-3_5

In this paper, we want to discuss the Riquier problem [20] for polyharmonic equations in $\mathbb{C}^m$. Our investigations in the higher dimensional case is based on the results we have obtained in the unit disc. In Sect. 2, we review the inhomogeneous Dirichlet problem in $\mathbb{C}$ for linear elliptic complex differential equations with polyharmonic leading term together with the relevant class of integral operators. In the next section we extend the properties in $\mathbb{C}$ into $\mathbb{C}^m$. The last two sections are devoted to the solution of the Riquier problem in $\mathbb{C}^2$.

2 Preliminaries

In this article we discuss the Riquier problem with a generalized polyharmonic operator as leading term in a polydisc in $\mathbb{C}^m$. We will extend the results obtained [1] in $\mathbb{C}$ to higher-dimensional case. Thus we start with providing a short review of the relevant information. In the unit disc $\mathbb{D}$ of the complex plane, the harmonic Green function is defined as

$$G_1(z,\zeta) = \log\left|\frac{1-z\bar{\zeta}}{\zeta - z}\right|^2 .$$

The properties of the harmonic Green function are well-known [5]. Since the relevant differential operator is self-adjoint, Green's function is symmetric, i.e., $G_1(z,\zeta) = G_1(\zeta,z)$ holds, [7]. $G_1(z,\zeta)$ is related to the Dirichlet problem for Poisson equation.

A polyharmonic Green function G_n is given iteratively by

$$G_n(z,\zeta) = -\frac{1}{\pi}\iint_{\mathbb{D}} G_1(z,\tilde{\zeta})G_{n-1}(\tilde{\zeta},\zeta)d\tilde{\xi}d\tilde{\eta}$$

for $n \geq 2$. It has the properties [11]

- $G_n(z,\zeta)$ is polyharmonic of order n in $\mathbb{D}\backslash\{\zeta\}$ for any $\zeta \in \mathbb{D}$,
- $G_n(z,\zeta) + \dfrac{|\zeta - z|^{2(n-1)}}{(n-1)!^2}\log|\zeta - z|^2$ is polyharmonic of order n for $z \in \mathbb{D}$ for any $\zeta \in \mathbb{D}$,
- $(\partial_z\partial_{\bar{z}})^{\mu}G_n(z,\zeta) = 0$ for $0 \leq \mu \leq n-1$ on $\partial\mathbb{D}$ for any $\zeta \in \mathbb{D}$,
- $G_n(z,\zeta) = G_n(\zeta,z)$, for any $z,\zeta \in \mathbb{D}$.

These functions are related to the n-Dirichlet problem for higher order Poisson equation [11] in the unit disc $\mathbb{D}$.

Theorem 2.1 ([11]) *The Riquier problem*

$$(\partial_z\partial_{\bar{z}})^n w = f \text{ in } \mathbb{D},\ (\partial_z\partial_{\bar{z}})^{\mu} w = \gamma_{\mu},\ 0 \leq \mu \leq n-1 \text{ on } \partial\mathbb{D}$$

with $f \in L^1(\mathbb{D}) \cap C(\mathbb{D}), \gamma_\mu \in C(\partial\mathbb{D}),\ 0 \le \mu \le n-1$ *is uniquely solvable. The solution is*

$$w(z) = -\sum_{\mu=1}^{n} \frac{1}{4\pi i} \int\limits_{\partial\mathbb{D}} \partial_{\nu_\zeta} G_\mu(z,\zeta)\gamma_{\mu-1}(\zeta)\frac{d\zeta}{\zeta} - \frac{1}{\pi}\iint_{\mathbb{D}} G_n(z,\zeta)f(\zeta)d\xi d\eta.$$

Now we define an integral operator related to Dirichlet type problems which is necessary to discuss the generalized n-Poisson equations in $\mathbb{C}$.

Definition 2.2 For $k, l \in \mathbb{N}_0$, $n \in \mathbb{N}$ $k + l \le 2n$, we define

$$G_n^{k,l} f(z) = -\frac{1}{\pi} \iint\limits_{\mathbb{D}} \partial_z^k \partial_{\bar z}^l G_n(z,\zeta) f(\zeta) d\xi d\eta$$

where $\zeta = \xi + i\eta$, for a suitable complex valued function f given in $\mathbb{D}$.

We need the properties in the sequel in order to discuss the existence and uniqueness of the solutions of generalized n-Poisson equation in $\mathbb{C}$. The main three deductions are stated in the following, which can be derived from results in [1] by slight changes in their proofs.

Lemma 2.3 *For* $k \in \mathbb{N}$, *if* $f \in W^{k,p}(\mathbb{D})$ *then*

$$\partial_z^{k-1} G_1^{2,0} f(z) = G_1^{1,0}((D - D_*)^k f(z)) \tag{2.1}$$

and $\partial_z^{k-1} G_1^{2,0}$ *is in* $L^p(\mathbb{D})$ *where* $Df(z) = \partial_z f(z)$, $D_* f(z) = \partial_{\bar z}(\bar z^2 f(z))$.

Corollary 2.4 *If*

$$f \in \begin{cases} L^p(\mathbb{D}) & , \ 1 \le k \le 2m-1 \\ W^{k+1-2m,p}(\mathbb{D}) & , \ k \ge 2m \end{cases}$$

then

$$G_n^{k,0} f(z) = G_1^{1,0}((D - D_*)^{k-1} G_n^{0,0} f(z)) \tag{2.2}$$

and $G_n^{k,0} f \in L^p(\mathbb{D})$ *hold.*

Theorem 2.5 *Let* $f \in L^p(\mathbb{D})$, $p > 2$ *and* $k + l < 2n$. *Then,*

$$\left|G_n^{k,l} f(z)\right| \le C\|f\|_{L^p(\mathbb{D})} \tag{2.3}$$

for $z \in \mathbb{D}$.

In the next section we extend these results to $\mathbb{C}^m$.

3 Riquier Problem in Higher Dimension

We may start with the representation formula defined for the variable $\hat{z}_k$ in $\mathbb{C}^m$

$$w(\hat{z}_k) = -\sum_{\mu=1}^{n} \frac{1}{4\pi i} \int_{\partial \mathbb{D}_k} \partial_{\nu_{\zeta_k}} G_\mu(\hat{z}_k, \hat{\zeta}_k)(\partial_{\hat{\zeta}_k} \partial_{\hat{\bar{\zeta}}_k})^{\mu-1} w(\hat{\zeta}_k) \frac{d\hat{\zeta}_k}{\hat{\zeta}_k} \tag{3.1}$$

$$-\frac{1}{\pi} \int_{\mathbb{D}_k} G_n(\hat{z}_k, \hat{\zeta}_k)(\partial_{\hat{\zeta}_k} \partial_{\hat{\bar{\zeta}}_k})^n w(\hat{\zeta}_k) d\xi_k d\eta_k.$$

We define the operators $\tilde{G}_{\mathbb{D}_k,n}$ and $\partial \tilde{G}_{\mathbb{D}_k,n}$ by

$$\tilde{G}_{\mathbb{D}_k,n}((\partial_{\hat{z}_k} \partial_{\hat{\bar{z}}_k})^n w(\hat{z}_k)) = -\frac{1}{\pi} \int_{\mathbb{D}_k} G_{k,n}(\hat{z}_k, \hat{\zeta}_k)(\partial_{\hat{\zeta}_k} \partial_{\hat{\bar{\zeta}}_k})^n w(\hat{\zeta}_k) d\xi_k d\eta_k \tag{3.2}$$

$$\partial \tilde{G}_{\mathbb{D}_k,n} w(\hat{z}_k) = -\sum_{\mu=1}^{n} \frac{1}{4\pi i} \int_{\partial \mathbb{D}_k} \partial_{\nu_{\zeta_k}} G_{k,\mu}(\hat{z}_k, \hat{\zeta}_k)(\partial_{\hat{\zeta}_k} \partial_{\hat{\bar{\zeta}}_k})^{\mu-1} w(\hat{\zeta}_k) \frac{d\hat{\zeta}_k}{\hat{\zeta}_k}. \tag{3.3}$$

where $G_{k,s}$ are Green's function of order s for the variable $\hat{z}_k$. Hence (3.1) can be written as

$$w(\hat{z}_k) = \partial \tilde{G}_{\mathbb{D}_k,n} w(\hat{z}_k) + \tilde{G}_{\mathbb{D}_k,n}((\partial_{z_k} \partial_{\bar{z}_k})^n w(\hat{z}_k)).$$

In two independent variable case, for example, we have

$$w(\hat{z}_1, z_2) = \partial \tilde{G}_{\mathbb{D}_1,n} w(\hat{z}_1, z_2) + \tilde{G}_{\mathbb{D}_1,n}((\partial_{z_1} \partial_{\bar{z}_1})^n w(\hat{z}_1, z_2)) \tag{3.4}$$

$$w(z_1, \hat{z}_2) = \partial \tilde{G}_{\mathbb{D}_2,n} w(z_1, \hat{z}_2) + \tilde{G}_{\mathbb{D}_2,n}((\partial_{z_2} \partial_{\bar{z}_2})^n w(z_1, \hat{z}_2)). \tag{3.5}$$

Substitute (3.5) in (3.4):

$$\begin{aligned}
w(z_1, z_2) &= \partial \tilde{G}_{\mathbb{D}_1,n}(\partial \tilde{G}_{\mathbb{D}_2,n} w + \tilde{G}_{\mathbb{D}_2,n}(\partial_{z_2} \partial_{\bar{z}_2})^n w) \\
&\quad + \tilde{G}_{\mathbb{D}_1,n}[(\partial_{z_1} \partial_{\bar{z}_1})^n (\partial \tilde{G}_{\mathbb{D}_2,n} w + \tilde{G}_{\mathbb{D}_2,n}(\partial_{z_2} \partial_{\bar{z}_2})^n w)] \\
&= \partial \tilde{G}_{\mathbb{D}_1,n}(\partial \tilde{G}_{\mathbb{D}_2,n} w) + \partial \tilde{G}_{\mathbb{D}_1,n}(\tilde{G}_{\mathbb{D}_2,n}(\partial_{z_2} \partial_{\bar{z}_2})^n w) \\
&\quad + \tilde{G}_{\mathbb{D}_1,n}((\partial_{z_1} \partial_{\bar{z}_1})^n (\partial \tilde{G}_{\mathbb{D}_2,n} w) + \tilde{G}_{\mathbb{D}_1,n}((\partial_{z_1} \partial_{\bar{z}_1})^n (\tilde{G}_{\mathbb{D}_2,n}(\partial_{z_1} \partial_{\bar{z}_1})^n w)) \\
&= \partial \tilde{G}_{\mathbb{D}^2,n} w + \partial \tilde{G}_{\mathbb{D}_1,n}(w - \partial \tilde{G}_{\mathbb{D}_2,n} w) \\
&\quad + \partial \tilde{G}_{\mathbb{D}_2,n}(\tilde{G}_{\mathbb{D}_1,n}(\partial_{z_1} \partial_{\bar{z}_1})^n w) + \tilde{G}_{\mathbb{D}^2,n}[(\partial_{z_1} \partial_{\bar{z}_1})^n (\partial_{z_2} \partial_{\bar{z}_2})^n w]
\end{aligned}$$

$$
\begin{aligned}
&= \partial\tilde{G}_{\mathbb{D}^2,n}w + \partial\tilde{G}_{\mathbb{D}_1,n}w - \partial\tilde{G}_{\mathbb{D}^2,n}w \\
&+ \partial\tilde{G}_{\mathbb{D}_2,n}(w - \partial\tilde{G}_{\mathbb{D}_1,n}w) + \tilde{G}_{\mathbb{D}^2,n}[(\partial_{z_1}\partial_{\bar{z_1}})^n(\partial_{z_2}\partial_{\bar{z_2}})^n w] \\
&= \partial\tilde{G}_{\mathbb{D}_1,n}w + \partial\tilde{G}_{\mathbb{D}_2,n}w - \partial\tilde{G}_{\mathbb{D}^2,n}w + \tilde{G}_{\mathbb{D}^2,n}[(\partial_{z_1}\partial_{\bar{z_1}})^n(\partial_{z_2}\partial_{\bar{z_2}})^n w] \qquad (3.6)
\end{aligned}
$$

where

$$
\tilde{G}_{\mathbb{D}^2,n}[(\partial_{z_1}\partial_{\bar{z_1}})^n(\partial_{z_2}\partial_{\bar{z_2}})^n w(z_1,z_2)]
$$

$$
\begin{aligned}
&= -\frac{1}{\pi}\int_{\mathbb{D}_1} G_{\mathbb{D}_1,n}(z_1,\zeta_1)(\partial_{\zeta_1}\partial_{\bar{\zeta_1}})^n\left[-\frac{1}{\pi}\int_{\mathbb{D}_2} G_{\mathbb{D}_2,n}(z_2,\zeta_2)(\partial_{\zeta_2}\partial_{\bar{\zeta_2}})^n w d\xi_2 d\eta_2\right] d\xi_1 d\eta_1 \\
&= \left(-\frac{1}{\pi}\right)^2\int_{\mathbb{D}_1}\int_{\mathbb{D}_2} G_{\mathbb{D}_1,n}(z_1,\zeta_1)G_{\mathbb{D}_2,n}(z_2,\zeta_2)(\partial_{\zeta_1}\partial_{\bar{\zeta_1}})^n(\partial_{\zeta_2}\partial_{\bar{\zeta_2}})^n w(\zeta_1,\zeta_2)d\xi_2 d\eta_2 d\xi_1 d\eta_1.
\end{aligned}
$$

We define

$$
G_{\mathbb{D}^2,n}(z_1,z_2;\zeta_1,\zeta_2) := G_{\mathbb{D}_1,n}(z_1,\zeta_1)G_{\mathbb{D}_2,n}(z_2,\zeta_2)
$$

as a Green's function for the polydisc $\mathbb{D}^2 := \mathbb{D}_1 \times \mathbb{D}_2$.

For a similar result in $\mathbb{C}^3$ we take

$$
w(\hat{z}_1,z_2,z_3) = \partial\tilde{G}_{\mathbb{D}_1,n}w(\hat{z}_1,z_2,z_3) + \tilde{G}_{\mathbb{D}_1,n}((\partial_{z_1}\partial_{\bar{z_1}})^n w(\hat{z}_1,z_2,z_3)) \qquad (3.7)
$$

$$
w(z_1,\hat{z}_2,z_3) = \partial\tilde{G}_{\mathbb{D}_2,n}w(z_1,\hat{z}_2,z_3) + \tilde{G}_{\mathbb{D}_2,n}((\partial_{z_2}\partial_{\bar{z_2}})^n w(z_1,\hat{z}_2,z_3)). \qquad (3.8)
$$

$$
w(z_1,z_2,\hat{z}_3) = \partial\tilde{G}_{\mathbb{D}_3,n}w(z_1,z_2,\hat{z}_3) + \tilde{G}_{\mathbb{D}_3,n}((\partial_{z_3}\partial_{\bar{z_3}})^n w(z_1,z_2,\hat{z}_3)). \qquad (3.9)
$$

Using the above equations we find

$$
\begin{aligned}
w(z_1,z_2,z_3) &= \partial\tilde{G}_{\mathbb{D}_1,n}w + \partial\tilde{G}_{\mathbb{D}_2,n}w - \partial\tilde{G}_{\mathbb{D}_1\times\mathbb{D}_2,n}w \\
&+ \partial\tilde{G}_{\mathbb{D}_3,n}w - \partial\tilde{G}_{\mathbb{D}_1\times\mathbb{D}_3,n}w - \partial\tilde{G}_{\mathbb{D}_2\times\mathbb{D}_3,n}w + \partial\tilde{G}_{\mathbb{D}^3,n}w \\
&+ \tilde{G}_{\mathbb{D}^3,n}[\Pi_{k=1}^3(\partial_{z_k}\partial_{\bar{z_k}})^n w].
\end{aligned}
$$

In $\mathbb{C}^m$ the representation will be

$$
w(z_1,\ldots,z_m) = \sum_{s=1}^{m}(-1)^{s+1}\sum_{j_1+j_2+\cdots+j_r=s}\partial\tilde{G}_{D_{j_1}\times\cdots\times D_{j_r},n}w + \tilde{G}_{\mathbb{D}^m,n}[\prod_{k=1}^{m}(\partial_{z_k}\partial_{\bar{z_k}})^n w]
$$

where

$$\tilde{G}_{\mathbb{D}^m,n}w = \left(-\frac{1}{\pi}\right)^m \int_{\mathbb{D}_1} \cdots \int_{\mathbb{D}_m} G_{\mathbb{D}^m,n}(z_1, \ldots, z_m; \zeta_1, \ldots, \zeta_m) w(\zeta) d\xi_1 d\eta_1 \ldots d\xi_m d\eta_m$$

in which

$$G_{\mathbb{D}^m,n}(z_1, \ldots, z_m; \zeta_1, \ldots, \zeta_m) = \prod_{j=1}^{m} G_{\mathbb{D}_j,n}(z_j, \zeta_j)$$

and

$$\partial \tilde{G}_{\mathbb{D}_{j_1} \times \cdots \times \mathbb{D}_{j_r},n} w = \partial \tilde{G}_{\mathbb{D}_{j_1} \times \cdots \times \mathbb{D}_{j_{r-1}},n}(\partial \tilde{G}_{\mathbb{D}_{j_r},n} w) \tag{3.10}$$

For the evaluation of (3.10), we will start with $\partial \tilde{G}_{\mathbb{D}_1}(\partial \tilde{G}_{\mathbb{D}_2} w)$:

$$\begin{aligned}
\partial \tilde{G}_{\mathbb{D}^2,n} w &= \partial \tilde{G}_{\mathbb{D}_1,n}(\partial \tilde{G}_{\mathbb{D}_2,n} w) \\
&= -\sum_{\mu=1}^{n} \frac{1}{4\pi i} \int_{\partial \mathbb{D}_1} \partial_{\nu_{\zeta_1}} G_{\mathbb{D}_1,\mu}(z_1, \zeta_1)(\partial_{\zeta_1} \partial_{\bar{\zeta}_1})^{\mu-1} (\partial G_{\mathbb{D}_2,n} w) \frac{d\zeta_1}{\zeta_1} \\
&= -\sum_{\mu=1}^{n} \frac{1}{4\pi i} \int_{\partial \mathbb{D}_1} \partial_{\nu_{\zeta_1}} G_{\mathbb{D}_1,\mu}(z_1, \zeta_1)(\partial_{\zeta_1} \partial_{\bar{\zeta}_1})^{\mu-1} \\
&\times \left[-\sum_{\alpha=1}^{n} \frac{1}{4\pi i} \int_{\partial \mathbb{D}_1} \partial_{\nu_{\zeta_2}} G_{\mathbb{D}_2,\alpha}(z_2, \zeta_2)(\partial_{\zeta_2} \partial_{\bar{\zeta}_2})^{\alpha-1} w(\zeta_1, \zeta_2) \frac{d\zeta_2}{\zeta_2} \right] \frac{d\zeta_1}{\zeta_1} \\
&= \left(-\frac{1}{4\pi i}\right)^2 \sum_{\mu=1}^{n} \sum_{\alpha=1}^{n} \int_{\partial \mathbb{D}^2} \partial_{\nu_{\zeta_1}} G_{\mathbb{D}_1,\mu}(z_1, \zeta_1) \partial_{\nu_{\zeta_2}} G_{\mathbb{D}_2,\alpha}(z_2, \zeta_2) \\
&\times (\partial_{\zeta_1} \partial_{\bar{\zeta}_1})^{\mu-1} (\partial_{\zeta_2} \partial_{\bar{\zeta}_2})^{\alpha-1} w(\zeta_1, \zeta_2) \frac{d\zeta_2}{\zeta_2} \frac{d\zeta_1}{\zeta_1} \\
&= \left(\frac{i}{4\pi}\right)^2 \int_{\partial \mathbb{D}^2} \sum_{\mu_1,\mu_2=1}^{n} \partial_{\nu_{\zeta_1}} \partial_{\nu_{\zeta_2}} G_{\mathbb{D}_1,\mu_1}(z_1, \zeta_1) G_{\mathbb{D}_2,\mu_2}(z_2, \zeta_2) \\
&\times (\partial_{\zeta_1} \partial_{\bar{\zeta}_1})^{\mu_2-1} (\partial_{\zeta_2} \partial_{\bar{\zeta}_2})^{\mu_2-1} w(\zeta_1, \zeta_2) \frac{d\zeta_2}{\zeta_2} \frac{d\zeta_1}{\zeta_1}.
\end{aligned}$$

It is easy to derive

$$\partial\tilde{G}_{\mathbb{D}_{j1}\times\cdots\times\mathbb{D}_{jr},n}w(z_1, z_2 \ldots, z_m) =$$

$$= \left(\frac{i}{4\pi}\right)^r \int\limits_{\partial\mathbb{D}_{j1}\times\cdots\times\partial\mathbb{D}_{jr},n} \sum_{\mu_{j1},\ldots,\mu_{jr}=1}^{n} \partial_{\nu_{\zeta_{j1}}} \ldots \partial_{\nu_{\zeta_{jr}}} G_{\mathbb{D}_{j1},\mu}(z_{j1}, \zeta_{j1}) \ldots G_{\mathbb{D}_{jr},\mu_{jr}}(z_{jr}, \zeta_{jr})$$

$$\times\ (\partial_{\zeta_{j1}} \partial_{\bar{\zeta}_{j1}})^{\mu_{j1}-1} \ldots (\partial_{\zeta_{jr}} \partial_{\bar{\zeta}_{jr}})^{\mu_{jr}-1} w(\zeta_1, \ldots \zeta_n) \frac{d\zeta_{jr}}{\zeta_{jr}} \ldots \frac{d\zeta_{j1}}{\zeta_{j1}}.$$

using mathematical induction. Combining the above results we may state an integral representation for a function $w \in \mathbb{C}^{2n}$.

Theorem 3.1 *Any function in* $\mathbb{C}^{2n}(\overline{\mathbb{D}}^m)$ *has the representation*

$$w(z_1, \ldots, z_m) =$$

$$= \left(\frac{i}{4\pi}\right)^r \int\limits_{\partial\mathbb{D}_{j1}\times\cdots\times\partial\mathbb{D}_{jr},n} \sum_{\mu_{j1},\ldots,\mu_{jr}=1}^{n} \partial_{\nu_{\zeta_{j1}}} \ldots \partial_{\nu_{\zeta_{jr}}} G_{\mathbb{D}_{j1},\mu}(z_{j1}, \zeta_{j1}) \ldots G_{\mathbb{D}_{jr},\mu_{jr}}(z_{jr}, \zeta_{jr})$$

$$\times\ (\partial_{\zeta_{j1}} \partial_{\bar{\zeta}_{j1}})^{\mu_{j1}-1} \ldots (\partial_{\zeta_{jr}} \partial_{\bar{\zeta}_{jr}})^{\mu_{jr}-1} w(\zeta_1, \ldots \zeta_n) \frac{d\zeta_{jr}}{\zeta_{jr}} \ldots \frac{d\zeta_{j1}}{\zeta_{j1}}$$

$$+ \left(-\frac{1}{\pi}\right)^m \int_{\mathbb{D}_1} \cdots \int_{\mathbb{D}_m} \prod_{j=1}^{m} G_{\mathbb{D}_j,n}(z_j, \zeta_j) \prod_{j=1}^{m} (\partial_{\zeta_j} \partial_{\bar{\zeta}_j})^n w(\zeta_1, \ldots \zeta_n) d\xi_1 d\eta_1 \ldots d\xi_m d\eta_m$$

4 Riquier Problems for Higher Order Polyharmonic Equations in a Polydisc

Now we may state a Riquier problem in $\mathbb{D}^m \subset \mathbb{C}^m$ for higher order polyharmonic equations:

Definition 4.1 Find a function $w \in W^{2nm,p}(\mathbb{D}^m)$ satisfying

$$(\partial_{z_1} \partial_{\bar{z}_1})^n (\partial_{z_2} \partial_{\bar{z}_2})^n \ldots (\partial_{z_m} \partial_{\bar{z}_m})^n w = f \ \ in \ \ \mathbb{D}^m$$

$$(\partial_{z_1} \partial_{\bar{z}_1})^{s_1} (\partial_{z_2} \partial_{\bar{z}_2})^{s_2} \ldots (\partial_{z_m} \partial_{\bar{z}_m})^{s_m} w = \gamma_{s_1,\ldots s_m}(z_1, \ldots, z_m)$$

for all $0 \leq s_j < n$ on $\partial\mathbb{D}^m$ where $\gamma_{s_1,\ldots s_m}$ are the restrictions of a polyharmonic function on the distinguished boundary for each s_j, is called a Riquier problem on $\mathbb{D}^m$.

In the rest of the article we concentrate on Dirichlet problems in $\mathbb{C}^2$. The integral representation of a function with the help of Green's function is given by (3.6) as

$$\begin{aligned} w(z_1, z_2) &= \partial\tilde{G}_{\mathbb{D}_1,n}w + \partial\tilde{G}_{\mathbb{D}_2,n}w - \partial\tilde{G}_{\mathbb{D}^2,n}w \\ &\quad + \tilde{G}_{\mathbb{D}^2,n}[(\partial_{z_1}\partial_{\bar{z}_1})^n(\partial_{z_2}\partial_{\bar{z}_2})^n w] \end{aligned}$$

where $\tilde{G}_{\mathbb{D}^2,n}$ and $\partial\tilde{G}_{\mathbb{D}_j,n}$ are defined by (3.2) and (3.3). Now we can state the following theorem:

Theorem 4.2 *The Riquier problem*

$$(\partial_{z_1}\partial_{\bar{z}_1})^n(\partial_{z_2}\partial_{\bar{z}_2})^n w = f(z_1, z_2) \;\; in \;\; \mathbb{D}^2 \tag{4.1}$$

subject to the conditions

$$(\partial_{z_1}\partial_{\bar{z}_1})^\mu(\partial_{z_2}\partial_{\bar{z}_2})^\nu w = \gamma_{\mu\nu}, \;\; 0 \le \mu, \nu \le n-1, \;\; on \;\; \partial\mathbb{D}^2 \tag{4.2}$$

is uniquely solvable for $f \in L^p(\mathbb{D}^2)$, $p \ge 2$, $\gamma_{\mu\nu} \in C^{4n-4}(\mathbb{D}^2)$ by

$$\begin{aligned} w(z_1, z_2) &= \sum_{\mu=1}^{n} \frac{i}{4\pi} \int_{\partial\mathbb{D}_1} \partial_{\nu_{\zeta_1}} G_{1,\mu}(z_1, \zeta_1)(\partial_{z_1}\partial_{\bar{z}_1})^{\mu-1}\gamma_{\mu,0}(\zeta_1, 0)\frac{d\zeta_1}{\zeta_1} \\ &+ \sum_{\mu=1}^{n} \frac{i}{4\pi} \int_{\partial\mathbb{D}_2} \partial_{\nu_{\zeta_2}} G_{2,\mu}(z_2, \zeta_2)(\partial_{z_2}\partial_{\bar{z}_2})^{\mu-1}\gamma_{0,\mu}(0, \zeta_2)\frac{d\zeta_2}{\zeta_2} \\ &+ \left(\frac{i}{4\pi}\right)^2 \int_{\partial\mathbb{D}^2} \sum_{\mu_1,\mu_2=1}^{n} \partial_{\nu_{\zeta_1}}\partial_{\nu_{\zeta_2}} G_{\mathbb{D}_1,\mu_1}(z_1, \zeta_1)G_{\mathbb{D}_2,\mu_2}(z_2, \zeta_2) \\ &\times (\partial_{\zeta_1}\partial_{\bar{\zeta}_1})^{\mu_1-1}(\partial_{\zeta_2}\partial_{\bar{\zeta}_2})^{\mu_2-1} w(\zeta_1, \zeta_2)\frac{d\zeta_2}{\zeta_2}\frac{d\zeta_1}{\zeta_1} \\ &+ \left(-\frac{1}{\pi}\right)^2 \int_{\mathbb{D}_1}\int_{\mathbb{D}_2} G_{\mathbb{D}_1,n}(z_1, \zeta_1)G_{\mathbb{D}_2,n}(z_2, \zeta_2)f(\zeta_1, \zeta_2)d\xi_2 d\eta_2 d\xi_1 d\eta_1. \end{aligned}$$

In the case of Riquier problem with homogeneous boundary conditions we get

$$w(z_1, z_2) = \frac{1}{\pi^2}\int_{\mathbb{D}_1}\int_{\mathbb{D}_2} G_{\mathbb{D}_1,n}(z_1, \zeta_1)G_{\mathbb{D}_2,n}(z_2, \zeta_2)f(\zeta_1, \zeta_2)d\xi_2 d\eta_2 d\xi_1 d\eta_1$$

or

$$w(z_1, z_2) = \frac{1}{\pi^2}\int_{\mathbb{D}_1}\int_{\mathbb{D}_2} G_{\mathbb{D}^2,n}(z_1, z_2, \zeta_1, \zeta_2)f(\zeta_1, \zeta_2)d\xi_2 d\eta_2 d\xi_1 d\eta_1$$

and hence

$$w(z_1, z_2) = \tilde{G}_{\mathbb{D}^2,n}f(z_1, z_2).$$

4.1 The Properties of Integral Operators Related to $\tilde{G}_{\mathbb{D}^2,n}$

Our aim in this article is to discuss the existence and uniqueness of the solutions of generalized linear higher order differential equations in $\mathbb{C}^2$. Now we define the relevant operators. Let us use the notations $\partial := (\partial_{z_1}, \partial_{z_2})$, $\overline{\partial} := (\partial_{\bar{z}_1}, \partial_{\bar{z}_2})$ and $\alpha = (\alpha_1, \alpha_2)$, $\beta = (\beta_1, \beta_2)$ where $\alpha_i, \beta_i \in \mathbb{N}$.

Definition 4.3 For $n \in \mathbb{N}$ and $|\alpha| \leq 2n$, $|\beta| \leq 2n$, we define

$$\partial^{\alpha}\overline{\partial}^{\beta}\tilde{G}_{\mathbb{D}^2,n}f(z_1, z_2) = -\frac{1}{\pi^2}\int\limits_{\mathbb{D}^2}\partial^{\alpha}\overline{\partial}^{\beta}G_{\mathbb{D}^2,n}(z_1, z_2; \zeta_1, \zeta_2)f(\zeta_1, \zeta_2)d\xi_1 d\eta_1 d\xi_2 d\eta_2$$

for a suitable complex valued function f defined in $\mathbb{D}^2$.

Lemma 4.4 *For $k, s \in \mathbb{N}$, if $f \in W^{k+s,p}(\mathbb{D})$ then*

$$\partial_{z_1}^{k+1}\partial_{z_2}^{s+1}\tilde{G}_{\mathbb{D}^2,1}f(z_1, z_2) = \partial_{z_1}\partial_{z_2}\tilde{G}_{\mathbb{D}^2,1}((D_1 - D_{1*})^k(D_2 - D_{2*})^s f(z_1, z_2)) \tag{4.3}$$

and $\partial_{z_1}^{k+1}\partial_{z_2}^{s+1}\tilde{G}_{\mathbb{D}^2,1}f(z_1, z_2)$ is in $L^p(\mathbb{D}^2)$ where $D_j f(z_1, z_2) = \partial_{z_j}f(z_1, z_2)$, $D_{j*}f(z_1, z_2) = \partial_{\bar{z}_j}(\bar{z}_j^2 f(z_1, z_2))$, $j = 1, 2$.

Proof Previously (in[1] Lemma 4.5) we had proved in $\mathbb{C}$ that

$$\partial_z^{k-1}G_1^{2,0}f(z) = G_1^{1,0}((D - D_*)^k f(z))$$

and $\partial_z^{k-1}G_1^{2,0}$ is in $L^p(\mathbb{D})$ where $Df(z) = \partial_z f(z)$, $D_* f(z) = \partial_{\bar{z}}(\bar{z}^2 f(z))$. Now we use this technique iteratively:

$$\begin{aligned}
&\partial_{z_1}^{k+1}\partial_{z_2}^{s+1}\tilde{G}_{\mathbb{D}^2,1}f(z_1, z_2) \\
&= \frac{1}{\pi^2}\partial_{z_1}^{k+1}\partial_{z_2}^{s+1}\int_{\mathbb{D}_1}\int_{\mathbb{D}_2}G_{\mathbb{D}_1,1}(z_1, \zeta_1)G_{\mathbb{D}_2,1}(z_2, \zeta_2)f(\zeta_1, \zeta_2)d\xi_2 d\eta_2 d\xi_1 d\eta_1 \\
&= \frac{1}{\pi^2}\partial_{z_1}^{k+1}\int_{\mathbb{D}_1}G_{\mathbb{D}_1,1}(z_1, \zeta_1)\left[\partial_{z_2}^{s+1}\int_{\mathbb{D}_2}G_{\mathbb{D}_2,1}(z_2, \zeta_2)f(\zeta_1, \zeta_2)d\xi_2 d\eta_2\right]d\xi_1 d\eta_1 \\
&= \frac{1}{\pi^2}\partial_{z_1}\partial_{z_2}\int_{\mathbb{D}_1}\int_{\mathbb{D}_2}G_{\mathbb{D}_1,1}(z_1, \zeta_1)G_{\mathbb{D}_2,1}(z_2, \zeta_2) \\
&\times ((D_1 - D_{1*})^k(D_2 - D_{2*})^s f(\zeta_1, \zeta_2))d\xi_2 d\eta_2 d\xi_1 d\eta_1
\end{aligned}$$

gives the required result. □

Corollary 4.5 *If*

$$f \in \begin{cases} L^p(\mathbb{D}^2) & , \ 1 \le k+s \le 4n-1 \\ W^{k+s+1-4n,p}(\mathbb{D}^2) \ , & k+s \ge 4n \end{cases}$$

then

$$\partial_{z_1}^k \partial_{z_2}^s \tilde{G}_{\mathbb{D}^2,n} f(z_1,z_2) = \partial_{z_1} \partial_{z_2} \tilde{G}_{\mathbb{D}^2,1}[(D_1 - D_{1*})^{k-1}(D_2 - D_{2*})^{s-1} \tilde{G}_{\mathbb{D}^2,n-1} f(z_1,z_2)] \tag{4.4}$$

and $\partial_{z_1}^k \partial_{z_2}^s \tilde{G}_{\mathbb{D}^2,n} f \in L^p(\mathbb{D}^2)$ *hold.*

Proof Let us recall that

$$\tilde{G}_{\mathbb{D}^2,n} f(z_1,z_2) = \tilde{G}_{\mathbb{D}^2,1}(\tilde{G}_{\mathbb{D}^2,n-1} f(z_1,z_2)).$$

Then by Lemma 4.4 we have

$$\begin{aligned} \partial_{z_1}^k \partial_{z_2}^s \tilde{G}_{\mathbb{D}^2,n} f(z_1,z_2) &= \partial_{z_1}^k \partial_{z_2}^s \tilde{G}_{\mathbb{D}^2,1}(\tilde{G}_{\mathbb{D}^2,n-1} f(z_1,z_2)) \\ &= \partial_{z_1} \partial_{z_2} \tilde{G}_{\mathbb{D}^2,1}[(D_1 - D_{1*})^{k-1}(D_2 - D_{2*})^{s-1} \tilde{G}_{\mathbb{D}^2,n-1} f(z_1,z_2)] \end{aligned}$$

and $\partial_{z_1}^k \partial_{z_2}^s \tilde{G}_{\mathbb{D}^2,n} f \in L^p(\mathbb{D}^2)$. This completes the proof. □

Note It is easy to show that

$$|\partial_{z_1}^{k+1} \partial_{z_2}^{s+1} \tilde{G}_{\mathbb{D}^2,n} f| \le C \|f\|_{L^p(\mathbb{D}^2)}$$

for $k+s < 4n$ and $p > 2$.

5 Riquier Problem for Generalized Higher Order Equations in $\mathbb{D}^2$

In this section we consider the equation

$$\partial^{(n,n)} \overline{\partial}^{(n,n)} w + \sum_{|\alpha|+|\beta| \le 4n} \left[q_{\alpha\beta}^{(1)}(z_1,z_2) \partial^{\alpha} \overline{\partial}^{\beta} w + q_{\alpha\beta}^{(2)}(z_1,z_2) \partial^{\beta} \overline{\partial}^{\alpha} \overline{w} \right] = f \ \ in\ \mathbb{D}^2 \tag{5.1}$$

where $q_{\alpha\beta}^{(1)}, q_{\alpha\beta}^{(2)}$ are measurable bounded functions and $f \in L^p(\mathbb{D}^2)$.

We take the Riquier problem for (5.1) subject to the homogeneous boundary conditions. In order to state a weak form of the existence and uniqueness, we assume that the solution be $w = \tilde{G}_{\mathbb{D}^2,n} g$ for some function g. Let us convert the given boundary value problem into an integral equation of the form

$$(I + B)g = f$$

where

$$Bg = \sum_{|\alpha|+|\beta|\leq 4n} \left[q^{(1)}_{\alpha\beta}(z_1, z_2)\partial^\alpha \overline{\partial}^\beta g + q^{(2)}_{\alpha\beta}(z_1, z_2)\partial^\beta \overline{\partial}^\alpha \bar{g} \right]. \tag{5.2}$$

If

$$q_0 \max_{|\alpha|+|\beta|\leq 4n, |\alpha|\neq|\beta|} \|\partial^\alpha \overline{\partial}^\beta \tilde{G}_{\mathbb{D}^2,n}\|_{L^p(\mathbb{D}^2)} < 1$$

where

$$\sum_{|\alpha|+|\beta|\leq 4n} |q^{(1)}_{\alpha\beta}(z_1, z_2)| + |q^{(2)}_{\alpha\beta}(z_1, z_2)| \leq q_0 < 1$$

then $\|B\|_{L^p(\mathbb{D}^2)} < 1$. Hence, $I + B$ is invertible [17] and we can use Fredholm alternative.Thus we can summarize this discussion as:

Theorem 5.1 *The Eq.* (5.1) *with homogeneous boundary conditions is solvable if*

$$\|B\|_{L^p(\mathbb{D}^2)} \leq q_0 \max_{|\alpha|+|\beta|\leq 4n, |\alpha|\neq|\beta|} \|\partial^\alpha \overline{\partial}^\beta \tilde{G}_{\mathbb{D}^2,n}\|_{L^p(\mathbb{D}^2)} < 1$$

and the solution is of the form $w = \tilde{G}_{\mathbb{D}^2,n} g$ *where* $g \in L^p(\mathbb{D}^2)$, $p > 2$, *is a solution of the integral equation* $(I + B)g = f$ *where*

$$Bg = \sum_{|\alpha|+|\beta|\leq 4n} \left[q^{(1)}_{\alpha\beta}(z_1, z_2)\partial^\alpha \overline{\partial}^\beta w + q^{(2)}_{\alpha\beta}(z_1, z_2)\partial^\beta \overline{\partial}^\alpha \bar{w} \right].$$

Acknowledgements The author is grateful to the anonymous referees for their careful reading which improved the article, and also to Professor Umit Aksoy for her valuable comments and supports.

References

1. Ü. Aksoy, A.O. Çelebi, Dirichlet problems for generalized n-Poisson equation, in *Pseudo-Differential Operators: Complex Analysis and Partial Differential Equations*, ed. by B.W. Schulze, M.W. Wong. Operator Theory: Advances and Applications, vol. 205 (Birkhäuser, Basel, 2009), pp. 129–141
2. Ü. Aksoy, A.O. Çelebi, Neumann problem for generalized n-Poisson equation. J. Math. Anal. Appl. **357**, 438–446 (2009)
3. Ü. Aksoy, A.O. Çelebi, Schwarz problem for higher order linear equations in a polydisc. Complex Var. Elliptic Equ. **62**, 1558–1569 (2016). https://doi.org/10.1080/17476933.2016.1254627
4. H. Begehr, Boundary value problems in complex analysis I,II, *Boletin de la Asosiación Matemática Venezolana*, vol. XII (2005), pp. 65–85, 217–250

5. H. Begehr, Biharmonic Green functions. Le Mathematiche **LXI**, 395–405 (2006)
6. H. Begehr, A particular polyharmonic Dirichlet problem, in *Complex Analysis Potential Theory*, ed. by T. Aliyev Azeroglu, T. M. Tamrazov (World Scientific Publishing, Singapore, 2007), pp. 84–115
7. H. Begehr, Six biharmonic Dirichlet problems in complex analysis, in *Function Spaces in Complex and Clifford Analysis* (National University Publishers, Hanoi, 2008), pp. 243–252
8. H.G.W. Begehr, A. Dzhuraev, *An Introduction to Several Complex Variables and Partial Differential Equations*. Pitman Monographs and Surveys in Pure and Applied Mathematics (Longman, Harlow, 1997)
9. H. Begehr, E. Gaertner, A Dirichlet problem for the inhomogeneous polyharmonic equation in the upper half plane. Georg. Math. J. **14**, 33–52 (2007)
10. H. Begehr, T. Vaitekhovich, Green functions in complex plane domains. Uzbek Math. J. **4**, 29–34 (2008)
11. H. Begehr, T. Vaitekhovich, Iterated Dirichlet problem for the higher order Poisson equation. Le Matematiche **LXIII**, 139–154 (2008)
12. H. Begehr, T.N.H. Vu, Z. Zhang, Polyharmonic Dirichlet problems. Proc. Steklov Inst. Math. **255**, 13–34 (2006)
13. H. Begehr, J. Du, Y. Wang, A Dirichlet problem for polyharmonic functions. Ann. Mat. Pure Appl. **187**, 435–457 (2008)
14. A. Krausz, Integraldarstellungen mit Greenschen Funktionen höherer Ordnung in Gebieten und Polygebieten (Integral representations with Green functions of higher order in domains and polydomains). Ph.D. thesis, FU Berlin, 2005. http://www.diss.fu-berlin.de/diss/receive/FUDISSthesis000000001659
15. A. Krausz, A general higher-order integral representation formula for polydomains. J. Appl. Funct. Anal. **2**(3), 197–222 (2007)
16. A. Kumar, A generalized Riemann boundary problem in two variables. Arch. Math. (Basel) **62**, 531–538 (1994)
17. S.S. Kutateladze, *Fundamentals of Functional Analysis* (Kluwer Academic Publishers, Dordrecht-Boston-London, 1996)
18. A. Mohammed, The Riemann-Hilbert problem for certain poly domains and its connection to the Riemann problem. J. Math. Anal. Appl. **343**, 706–723 (2008)
19. A. Mohammed, R. Schwarz, , Riemann-Hilbert problems and their connections in polydomains, in *Pseudo-Differential Operators: Complex Analysis and Partial Differential Equations*. Operator Theory: Advances and Applications, vol. 205 (Birkhäuser, Basel, 2010), pp. 143–166
20. M. Nicolescu, *Les functions polyharmonique* (Hermann, Paris, 1936)
21. T. Vaitsiakhovich, Boundary value problems for complex partial differential equations in a ring domain, Ph. D. thesis, F.U. Berlin, 2008
22. I.N. Vekua, *Generalized Analytic Functions* (Pergamon Press, Oxford, 1962)

A Microscopic Model of Redistribution of Individuals Inside an 'Elevator'

Marina Dolfin, Mirosław Lachowicz, and Andreas Schadschneider

Abstract We present and qualitatively analyze a stochastic microscopic model of redistribution of individuals inside a domain which can be thought as representing an elevator. The corresponding mesoscopic model is also derived.

Keywords Markov jump processes • Microscopic model • Individuals • Elevator

Mathematics Subject Classification (2010) Primary: 00A69, 35R09, 35Q91; Secondary: 91B72. 91B74, 39A30

1 Introduction

This paper presents a mathematical description at the microscopic level of the redistribution of individuals in a closed domain featuring, as an example, an elevator. Starting from the microscopic mathematical representation wc consider also the corresponding model at the mesoscopic level.

M. Dolfin
Dipartimento di Ingegneria, Universitá di Messina, Contrada Di Dio, Villaggio S. Agata, I–98166 Messina, Italy
e-mail: mdolfin@unime.it

M. Lachowicz (✉)
Dipartimento di Ingegneria, Universitá di Messina, Contrada Di Dio, Villaggio S. Agata, I–98166 Messina, Italy

Instytut Matematyki Stosowanej i Mechaniki, Uniwersytet Warszawski, ul. Banacha, 2, PL–02–097 Warsaw, Poland
e-mail: lachowic@mimuw.edu.pl

A. Schadschneider
Institut für Theoretische Physik, Universität zu Köln, Zülpicher Str., 77, D–50937 Köln, Germany
e-mail: as@thp.uni-koeln.de

P. Drygaś, S. Rogosin (eds.), *Modern Problems in Applied Analysis*,
Trends in Mathematics, https://doi.org/10.1007/978-3-319-72640-3_6

Regarding the topic of crowd dynamics, a huge literature can be found, with models that usually refer to very different scales: the macroscopic level ([18] and references therein), the mesoscopic level ([3] and references therein), the microscopic level ([13] and references therein) and even the nanoscale level [4]. Recently, a new kind of experiments have been performed in order to elucidate the interactions between pedestrians [9, 10, 16, 17]. In these experiments the inflow of persons into a spatially restricted area, e.g. an elevator, was studied, featuring the process inverse to evacuation.

The model here presented refers to the position of every separate individual (an agent of the system). We introduce two main classes of probabilities: one characterizing the influence of the domain, representing the wall and the entrance of the elevator, and another one characterizing the effect of the interactions among agents. The first one is due to two contributions, respectively describing the tendency to concentrate close to the entrance and the tendency to stay close to the boundary. The second one simply represents the tendency to be not too close to other agents. This last one is mainly related to the concept of personal space [1, 20] which is in turn strictly related to cultural and social mechanisms [12]. The relationship of an agent with the boundaries of a closed domain in which he/she is temporary located, reflects social influences too. The way in which individuals choose their temporary location is possibly related to the desire of avoiding social conflicts and misunderstanding but also to achieve the desired level of privacy and optimizing the position for exit. The relationship with the boundaries of such a closed domain like an elevator is also related to the phenomenon of crowding [1]. It is only indirectly related to the density, reflecting social influences like the relationship of the individuals with the neighbors.

The aim of the present paper is to propose a microscopic (individually-based) model (Sects. 2 and 4) that is able to describe the redistribution of individuals in a closed domain and to give preliminary analysis of this model (Sect. 5). The further detailed numerical simulations are left to the paper [5] in preparation.

We also propose the resulting mesoscopic model (Sect. 3). In order to test the applicability of the mesoscopic model to situations where the number of agents is not very large we will perform numerical simulations [5]. In any case, the mesoscopic model will provide relevant information about the system and might be considered as an approximated model.

2 General Framework: Microscopic Level

We introduce the general framework at the microscopic level, which will be adopted in our modelization, of a process of inflow and distribution of individuals inside an environment resembling the main features of an elevator. We consider a system of M ($< \infty$) agents (see [2, 14] and references therein). The n–th agent, with $n = 1, \ldots, M$, is characterized by its position

$$x_n \in \mathbb{U} \,,$$

outside $\mathbb{U}_{\text{out}}$ or inside $\mathbb{U}_{\text{in}}$ the 'elevator',

$$\mathbb{U}_{\text{out}} \cup \mathbb{U}_{\text{in}} = \mathbb{U}\,, \qquad \mathbb{U}_{\text{out}} \cap \mathbb{U}_{\text{in}} = \emptyset\,.$$

We assume that the evolution of the system is defined by the corresponding to M interacting agents (see [2, 6, 14] and references therein) evolution. We define the evolution of probability densities given by an evolution equation (the so-called modified Liouville equation) defined by a (linear) generator. The interactions are defined by the linear generator that completely describes the evolution of the probability density at the microscopic scale.

The n_1–agent changes its position at random times,

- a change occurs without any interaction—the agent is entering and choosing its position in the elevator or it is going out;
- a change due to the interaction with the $n_2, n_3, \ldots, n_M$ agents

$$n_2, \ldots, n_M \in \{1, \ldots, M\}\,.$$

The rate of change of position by the agent n_1 with position x_{n_1} entering the elevator and choosing the position or going out is given by a measurable function $a^{[1]} = a^{[1]}(x_{n_1})$ such that

$$0 \le a^{[1]}(x_{n_1}) \le a_+^{[1]} < \infty\,, \qquad \forall\, x_{n_1} \in \mathbb{U}\,, \tag{2.1}$$

where $a_+^{[1]}$ is a constant.

The corresponding transition is described by the measurable function

$$A^{[1]} = A^{[1]}\big(x; x_{n_1}\big) \ge 0\,, \qquad \forall\, x, x_{n_1} \in \mathbb{U}\,, \tag{2.2}$$

$A^{[1]}$ is a transition probability and therefore

$$\int_{\mathbb{U}} A^{[1]}\big(x\,;\, x_{n_1}\big)\, \mathrm{d}x = 1\,, \tag{2.3}$$

for all $x_{n_1} \in \mathbb{U}$ such that

$$a^{[1]}(\,x_{n_1}\,) > 0\,.$$

The rate of interaction between the agent n_1 with position x_{n_1} and the agents $n_2, \ldots, n_M$ with positions $x_{n_2}, \ldots, x_{n_M}$, is given by a measurable function $a^{[M]} = a^{[M]}(x_1, \ldots, x_M)$ such that

$$0 \le a^{[M]}(x_{n_1}, x_{n_2}, \ldots, x_{n_M}) \le a_+^{[M]} < \infty\,, \qquad \forall\, x_{n_1}, \ldots, x_{n_M} \in \mathbb{U}\,, \tag{2.4}$$

where $a_+^{[M]}$ is a constant.

The transition into position x of an agent n_1 with position x_{n_1}, due to the interaction with agents $n_2, \ldots, n_M$ with positions $x_{n_2}, \ldots, x_{n_M}$, respectively, is described by the measurable function

$$A^{[M]} = A^{[M]}\big(x; x_{n_1}, \ldots, x_{n_M}\big) \geq 0\,, \qquad \forall\, x, x_{n_1}, \ldots, x_{n_M} \in \mathbb{U}\,, \tag{2.5}$$

$A^{[M]}$ is a transition probability and therefore

$$\int_{\mathbb{U}} A^{[M]}\big(x\,;\, x_{n_1}, x_{n_2}, \, \ldots\, , x_{n_M}\big)\, \mathrm{d}x = 1\,, \tag{2.6}$$

for all $x_{n_1}\,, \, \ldots\, , x_{n_M} \in \mathbb{U}$ such that

$$a^{[M]}(\, x_{n_1}\,, \, \ldots\, , \, x_{n_M}\,) > 0\,.$$

The stochastic model (at the microscopic level) will be completely determined by the functions a and A. Different choices of a and A give rise to different microscopic models (Markov jump processes).

Given M, a and A, we assume that the stochastic system is defined by the Markov jump process of M agents through the following generator Λ

$$\begin{aligned}
&\Lambda\phi(x_1, \ldots, x_M) = \\
&= \sum_{1\leq n\leq M} a^{[1]}(x_n)\left(\int_{\mathbb{U}} A^{[1]}(x; x_n)\phi\big(x_1, \ldots, x_{n-1}, x, x_{n+1}, \ldots, x_M\big)\mathrm{d}x \right.\\
&\left. -\phi(x_1, \ldots, x_M)\right) + \\
&+ \sum_{\substack{1\leq n_1,\ldots,n_M\leq M\\ n_i\neq n_j\ \forall\, i\neq j}} a^{[M]}\big(x_{n_1}, x_{n_2}, \ldots, x_{n_M}\big)\left(\int_{\mathbb{U}} A^{[M]}\big(x; x_{n_1}, x_{n_2}, \ldots, x_{n_M}\big) \right.\\
&\left. \times\phi\big(x_1, \ldots, x_{n_1-1}, x, x_{n_1+1}, \ldots, x_M\big)\mathrm{d}x - \phi(x_1, \ldots, x_M)\right)\,.
\end{aligned}$$

Λ is the generator for a Markov jump process in $\mathbb{U}^M$ that can be constructed as in Ref. [7, Section 4.2]—cf. [14]. In this paper, as in [2, 6, 14], we refer to the evolution of probability densities. We assume now that the system is initially distributed according to the probability density $\overset{\circ}{f} \in L_1^{(M)}$, where $L_1^{(M)}$ is the space equipped with the norm

$$\|f\|_{L_1^{(M)}} = \int_{\mathbb{U}} \cdots \int_{\mathbb{U}} \big|f\big(x_1, \ldots, x_M\big)\big|\, \mathrm{d}x_1 \ldots \mathrm{d}x_M\,,$$

and the time evolution is described by the following linear equation

$$\frac{\partial}{\partial t} f = \Lambda^* f\,; \qquad f\Big|_{t=0} = \overset{\circ}{f}\,, \tag{2.7}$$

where Λ^* is the generator

$$\Lambda^* f\big(t, x_1, x_2, \ldots, x_M\big) =$$

$$+ \sum_{1 \le n \le M} \Bigg(\int_{\mathbb{U}} A^{[1]}\big(x_n; x\big) a^{[1]}\big(x\big) f\big(t, x_1, \ldots, x_{n-1}, x, x_{n+1}, \ldots, x_M\big)\, \mathrm{d}x$$

$$- a^{[1]}\big(x_n\big) f\big(t, x_1, \ldots, x_M\big) \Bigg) +$$

$$+ \sum_{\substack{1 \le n_1, \ldots, n_M \le M \\ n_i \ne n_j \ \forall\, i \ne j}} \Bigg(\int_{\mathbb{U}} A^{[M]}\big(x_{n_1}; x, x_{n_2}, \ldots, x_{n_M}\big) a^{[M]}\big(x, x_{n_2}, \ldots, x_{n_m}\big)$$

$$\times f\big(t, x_1, \ldots, x_{n_1 - 1}, x, x_{n_1 + 1}, \ldots, x_M\big)\, \mathrm{d}x$$

$$- a^{[M]}\big(x_{n_1}, x_{n_2}, \ldots, x_{n_m}\big) f\big(t, x_1, \ldots, x_M\big) \Bigg)\,.$$

The generator is the difference between the *gain term* and the *loss term*, where

- the *gain term* is the sum of terms describing the changes from the state x of the n_1–agent into x_{n_1} due to the interaction with the $n_2, \ldots, n_M$ agents with the states $x_{n_2}, \ldots, x_{n_M}$, respectively and the term describing the direct changes of the state x of the n_1-element into x_{n_1} without the interactions;
- the *loss term* is the sum of terms describing the changes from the state x_{n_1} of the n_1–agent into another state due to the interaction with the $n_2, \ldots, n_m$ agents with the states $x_2, \ldots, x_M$, respectively or without interactions.

Under Assumptions (2.4), (2.5), (2.6), the operator Λ^* is a bounded linear operator in the space $L_1^{(M)}$. Therefore the Cauchy Problem (2.7) has the unique solution given by a uniformly continuous semigroup according to the formula

$$f(t) = e^{t\Lambda^*} \overset{\circ}{f}$$

in $L_1^{(M)}$ for all $t \ge 0$. Moreover, by standard argument—cf. Refs. [2, 14]—we see that the solution is nonnegative for nonnegative initial data and the $L_1^{(M)}$-norm is preserved

$$\| f(t) \|_{L_1^{(M)}} = \| \overset{\circ}{f} \|_{L_1^{(M)}} = 1\,, \qquad \text{for } t > 0\,. \tag{2.8}$$

Hence, $\left(e^{t\Lambda^*}\right)_{t\geq 0}$ is a uniformly continuous semigroup of Markov operators; that is, a uniformly continuous stochastic semigroup.

3 Mesoscopic Level

Although the number of agents in applications cannot be a large number we may consider the corresponding mesoscopic model. The linear Eq. (2.7) may be related, [6, 14], with a nonlinear integro-differential equation corresponding to the mesoscopic description,

$$\frac{\partial}{\partial t} f(t,x) = \mathfrak{G}[f](t,x) - f(t,x)\mathfrak{L}[f](t,x), \qquad x \in \mathbb{U}, \tag{3.1}$$

where $\mathfrak{G}[f]$ is the *gain term*, given by

$$\begin{aligned}\mathfrak{G}[f](t,x) = & \int_{\mathbb{U}} A^{[1]}(x;y)a^{[1]}(y)f(t,y)\,\mathrm{d}y + \\ & + \sum_{\{\}} \int_{\mathbb{U}^M} A^{[M]}\big(x;x_1,\{x_2,\ldots,x_M\}\big)a^{[M]}\big(x_1,\{x_2,\ldots,x_M\}\big) \times \\ & \times f(t,x_1)f(t,x_2)\ldots f(t,x_M)\,\mathrm{d}x_1\,\mathrm{d}x_2\ldots\mathrm{d}x_M\,,\end{aligned}$$

and $f\mathfrak{L}[f]$ is the *loss term*, defined as

$$\begin{aligned}\mathfrak{L}[f](t,x) = & \; a^{[1]}(x) + \\ & + \sum_{\{\}} \int_{\mathbb{U}^{M-1}} a^{[M]}\big(x,\{x_2,\ldots x_M\}\big)f(t,x_2)\ldots f(t,x_M)\,\mathrm{d}x_2\ldots\mathrm{d}x_M\,,\end{aligned}$$

where $\sum_{\{\}}$ means the sum over all permutation of variables within { }.

4 Model of Redistribution Inside an 'Elevator'

First of all we introduce a short description of some experimental results regarding the inflow of individuals inside an "elevator" and their temporary distribution inside (the experiments artificially reproduce the real process of peoples entering an elevator [8, 10, 17]. The experiments we would like to focus on regard mostly the inflow process, such as in the case of individuals entering an elevator and choosing a temporary location inside. As remarked in [9–11] the inflow and the outflow process have very different main features. The basic hypotheses formulated on the basis of experimental studies [9, 10, 16, 17] are that individuals entering a closed space, like an elevator, prefer locations near to the boundary, mostly avoid the entrance area but at the same time they would like to minimize the cost of going out by concentrating near the entrance. It is worth to mention that the experimental results show a dependence on the density of individuals inside the domain. We would like to consider these psychological factors emerging from the experiments in choosing the specifications for the quantities a and A into Eq. (2.7). In this first step of the modelization, we assume that the number of persons M in the elevator is constant thus neglecting the inflow and going out; in this case is $\mathbb{U}_{\text{out}} = \emptyset$. Thus $\mathbb{U} = \mathbb{U}_{\text{in}}$. Moreover we may assume

$$a^{[1]} = \text{const} > 0\,, \tag{4.1}$$

and

$$a^{[M]} = \text{const} > 0\,. \tag{4.2}$$

Let the probability density $A^{[1]}(\,.\,, x_1)$ be independent on x_1. It expresses the probability of choosing the position without interactions with other agents. We assume that

$$A^{[1]}(x) = \lambda_1 A_1^{[1]}(x) + \lambda_2 A_2^{[1]}(x)\,, \qquad \lambda_i \geq 0 \quad \text{and} \quad \lambda_1 + \lambda_2 = 1\,, \tag{4.3}$$

where $A_1^{[1]}$ describes the tendency to concentrate close to the entrance and $A_2^{[1]}$ describes the tendency to be close to the boundary. $A_1^{[1]}$ and $A_2^{[1]}$ are probability densities on $\mathbb{U}$.

$A^{[M]}$ expresses the interactions among agents—the tendency to not be too close to other agents. Therefore $A^{[M]} = A^{[M]}(x; x_1, \ldots, x_M)$ is a probability density with respect to x variable (for all $x_1, x_2, \ldots, x_M$ in $\mathbb{U}$) on $\mathbb{U}$ such that

$$A^{[M]}(x; x_1, x_2 \ldots, x_M) = 0 \quad \text{on} \quad \mathbb{U}_{x_2,\ldots,x_M}\,, \tag{4.4}$$

where $\mathbb{U}_{x_2,\ldots,x_M} = \mathbb{U} \cap \bigcup_{i=2,\ldots,M} \mathbb{B}_i$, and $\mathbb{B}_i$ is a ball with the center at x_i and the small radius $\delta > 0$ such that $2M\pi\delta^2 < |\mathbb{U}|$. Moreover we assume that $A^{[M]} = A^{[M]}(x; x_1, \ldots, x_M)$ is a (quickly) decreasing function of $|x - x_1|$.

The specific assumptions on the quantities $A^{[1]}(x)$ and $A^{[M]}(x; x_1, \ldots, x_M)$ lead, together with Eq. (2.7) to the model of redistribution of individuals inside the 'elevator'.

5 Qualitative Analysis

We present a qualitative analysis of the proposed model leaving the simulations and comparison with the experiments for a forthcoming paper [5].

Consider the following assumption

Assumption 5.1 *Let at least one of the following two conditions be satisfied*

$$A^{[1]}\big(x_n; x\big)a^{[1]}\big(x\big) \geq \delta_1 > 0 \tag{5.1}$$

for all $x, x_n \in \mathbb{U}$;

or

$$A^{[M]}\big(x_{n_1}; x, x_{n_2}, \ldots, x_{n_M}\big)a^{[M]}\big(x, x_{n_2}, \ldots, x_{n_m}\big) \geq \delta_2 > 0 \tag{5.2}$$

for all $x_{n_1}, x_{n_2}, x \in \mathbb{U}$.

We note that, in the context of the present paper, Assumption (5.1) seems quite natural, whereas Assumption (5.2) is rather artificial. However in a simplified case, when $\mathbb{U}_{\text{out}} = \emptyset$ it may also be natural.

We can see that $\left(e^{t\Lambda^*}\right)_{t\geq 0}$ is partially integral—cf. [19]—and under Assumption 5.1 we have

Theorem 5.2 *Let the Assumptions* (2.1)–(2.6) *and Assumption* 5.1 *be satisfied. Then the semigroup* $\left(e^{t\Lambda^*}\right)_{t\geq 0}$ *is asymptotically stable.*

Proof Assume that Eq. (5.1) is satisfied. The proof for the case of Eq. (5.2) is analogous.

We rewrite Eq. (2.7) in the form

$$\frac{\mathrm{d}}{\mathrm{d}t} f = \Gamma f - a_+ f\,, \tag{5.3}$$

where $a_+ = Ma_+^{[1]} + M!a_+^{[M]}$ and Γ is a positive operator. Then

$$e^{t\,\Lambda^*} \overset{\circ}{f} = e^{-a_+ t} e^{t\,\Gamma} \overset{\circ}{f}\,, \tag{5.4}$$

The semigroup $e^{t\,\Lambda^*}$ is asymptotically stable iff the operator $e^{t_0\,\Lambda^*}$, for some $t_0 > 0$, is asymptotically stable as the operator defining a discrete dynamical system, cf. [15, 19]. We may consider e.g. $t_0 = 1$.

Let $n \geq 1$, we have

$$e^{n\,\Lambda^*} \overset{\circ}{f} = e^{\Lambda^*} e^{(n-1)\,\Lambda^*} \overset{\circ}{f}\,. \tag{5.5}$$

We note that $\hat{f} = e^{(n-1)\,\Lambda^*}\,\overset{\circ}{f}$ is a probability density. On the other hand, for any probability density $\hat{f}$,

$$e^{\Lambda^*}\hat{f} = e^{-a_+}e^{\Gamma}\hat{f} \geq e^{-a_+}\frac{1}{M!}\Gamma^M\hat{f} \geq \mathfrak{c}_M\,\|\hat{f}\|_M\,, \tag{5.6}$$

where $\mathfrak{c}_M$ is a positive (> 0) constant (that depends on M and a_+).

Because $\|\hat{f}\|_{L_1^{(M)}} = 1$ we obtain

$$e^{n\,\Lambda^*}\,\overset{\circ}{f} \geq \mathfrak{c}_M\,, \tag{5.7}$$

for any $n \geq 1$ and any probability density $\overset{\circ}{f}$.

Therefore a lower function for the semigroup $e^{t\Lambda^*}$ exists and the semigroup is stable—cf. [15, 19]. □

6 Conclusions

The study of inflow processes is less developed than that of outflow processes although the two phenomena may have strong correlations ([11] and references therein). Then, a better understanding of inflow processes may help in a deeper analysis of important processes related to individual safety, like evacuation dynamics. We presented a stochastic microscopic model and the related model at the mesoscopic level. A qualitative analysis of the general mathematical structure adopted has been given.

The numerical simulation in a work Ref. [5] in preparation will indicate the advantages and possible disadvantages of both approaches proposed in the present paper: the microscopic and the mesoscopic.

Acknowledgements This work was completed with the support of the university of Messina through the grant Visiting Professor 2016.

References

1. I. Altman, *The Environment and Social Behavior: Privacy, Personal Space, Territory, Crowding* (Brooks/Cole, Monterey, CA, 1975)
2. J. Banasiak, M. Lachowicz, *Methods of Small Parameter in Mathematical Biology* (Birkhäuser, Basel, 2014)
3. N. Bellomo, A. Bellouquid, D. Knopoff, From the microscale to collective crowd dynamics. Multiscale Model. Simul. **11**, 943–963 (2013)
4. E. Cristiani, B. Piccoli, A. Tosin (eds.), *Multiscale Modeling of Pedestrian Dynamics* (Springer, Berlin, 2014)

5. M. Dolfin, D. Knopoff, M. Lachowicz, A. Schadschneider, Monte Carlo simulation of space redistribution in an inflow process of individuals, in preparation
6. M. Dolfin, M. Lachowicz, Z. Szymańska, *A General Framework for Multiscale Modeling of Tumor – Immune System Interactions, In Mathematical Oncology*, ed. by A. d'Onofrio, A. Gandolfi (Birkhäuser, Basel, 2014), pp. 151–180
7. S.N. Ethier, T.G. Kurtz, *Markov Processes, Characterization and Convergence* (Wiley, New York, 1986)
8. T. Ezaki, D. Yanagisawa, K. Ohtsuka, K. Nishinari, Simulation of space acquisition process of pedestrians using proxemic floor field model. Physica A **391**, 291–299 (2012)
9. T. Ezaki, K. Ohtsuka, D. Yanagisawa, K. Nishinari, Inflow process: a counterpart of evacuation, in *Traffic and Granular Flow 2013*, ed. by M. Chraibi, M. Boltes, A. Schadschneider, A., Seyfried (Springer, Berlin, 2015), pp. 227–231
10. T. Ezaki, K. Ohtsuka, M. Chraibi, M. Boltes, D. Yanagisawa, A. Seyfried, A. Schadschneider, K. Nishinari, Inflow process of pedestrians to a confined space. Collective Dyn. **1**(A:4), 1–18 (2016)
11. K.A. Forche, Stochastic model of pedestrian inflow to a confined space, Mater's thesis, University of Cologne
12. E.T. Hall, *The Hidden Dimension* (Anchor Books, New York, 1966)
13. W. Klingsch, C. Rogsch, A. Schadschneider, M. Schreckenberg (eds.), *Pedestrian and Evacuation Dynamics 2008* (Springer, Berlin, 2010)
14. M. Lachowicz, Individually-based Markov processes modeling nonlinear systems in mathematical biology. Nonlinear Anal. Real World Appl. **12**, 2396–2407 (2011)
15. A. Lasota, J.A. Yorke, Exact dynamical systems and the Frobenius–Perron operator. Trans. Am. Math. Soc. **273**, 375–384 (1982)
16. X. Liu, W. Song, L. Fu, Z. Fang, Experimental study of pedestrian inflow in a room with a separate entrance and exit. Physica A **442**, 224–238 (2016)
17. X. Liu, W. Song, L. Fu, W. Lv, Z. Fang, Typical features of pedestrian spatial distribution in the inflow process. Phys. Lett. A **380**, 1526–1534 (2016)
18. M.D. Rosini, *Macroscopic Models for Vehicular Flows and Crowd Dynamics: Theory and Applications*. Springer Understanding Complex Systems (Springer, Heidelberg, 2013)
19. R. Rudnicki, *Models of population dynamics and their applications in genetics*, ed. by M. Lachowicz, J. Miękisz. From Genetics to Mathematics (World Scientific, Singapore, 2009), pp. 103–147
20. R. Sommer, Studies in personal space. Sociometry **22**, 247–260 (1959)

New Approach to Mathematical Model of Elastic in Two-Dimensional Composites

Piotr Drygaś

Abstract This paper is devoted to boundary value problems for elastic problems modelled by the biharmonic equation in two-dimensional composites. All the problems are studied via the method of complex potentials. The considered boundary value problems for analytic functions are reduced to functional-differential equations. Applications to calculation of the effective properties tensor are discussed.

Keywords Functional equation • Two-dimensional elastic composite • Eisenstein and Natanzon series • Effective stress properties

Mathematics Subject Classification (2010) Primary 30E25; Secondary 74Q15

1 Introduction

Many scientists consider two dimensional elastic composites with non-overlapping inclusions [1–4]. The most common approach is the numerical one. On the other hand, this problem can be discussed through boundary value problems for analytic functions following Muskhelishvili's approach [5]. The complex representation of the plane theory of the elasticity brought many new results. However, the authors using the analytic representation of the complex potentials apply integral methods [6–9]. Cited authors present separately methods for solving a local problem and an effective problem. We use a method of functional equations, which was proposed to solve the Riemann-Hilbert and $\mathbb{R}$-linear problems for multiply connected domains [10]. We reduce the problem under study for a circular multiply connected domain to a system of functional-differential equations and propose a constructive method for its solution. For the local problem we obtain exact solution by method of successive

P. Drygaś (✉)
Faculty of Mathematics and Natural Sciences, University of Rzeszów, Pigonia 1, 35-959 Rzeszów, Poland
e-mail: drygaspi@ur.edu.pl

P. Drygaś, S. Rogosin (eds.), *Modern Problems in Applied Analysis*,
Trends in Mathematics, https://doi.org/10.1007/978-3-319-72640-3_7

approximation, while for the effective problem we extend the local solution using the generalized Eisenstein and Natanzon functions [11, 12]. Independently, the same functions were introduced and applied to elastic problems by Filshtinsky (see [13–15]). This paper contain the results presented in the series of work [11, 16, 17].

2 Geometry and Functional Spaces

Consider n disks $D_k = \{z \in \mathbb{C} : |z - a_k| < r_k\}$, $(k = 1, .., n)$ in the complex plane $\mathbb{C}$. Let $\Gamma_k = \partial D_k$, $\widehat{C} = \mathbb{C} \cup \{\infty\}$, $\dot{D} = \mathbb{C} \setminus \bigcup_{j=1}^{n} \left(D_j \cup \Gamma_j\right)$, $D = \dot{D} \cup \{\infty\}$, where the curve Γ_k is counter clockwise orientated. The domain D is a multiply connected domain in $\mathbb{R}^2$. This geometry is equivalent to the composite containing unidirectional parallel fibres.

Let $z_{(k)}^*$ denote the inversion with respect to the circle Γ_k:

$$z_{(k)}^* := \frac{r_k^2}{z - a_k} + a_k, \qquad k = 1, \ldots, n.$$

Analytic functions considered in the present paper can be continuous and their second derivatives are continuous in the closures of the analyticity domains. For fixed $k = 1, \ldots, N$ the Hardy-Sobolev space $\mathcal{H}^{(2,2)}(D_k)$ [17] is a space of all functions f analytic in D_k, endowed with the norm

$$\|f\|^2_{\mathcal{H}^{(2,2)}(D_k)} := \sup_{0<r<r_k} \int_0^{2\pi} |f_j\left(re^{i\theta} + a_j\right)|^2 d\theta +$$
$$\sup_{0<r<r_k} \int_0^{2\pi} |f_j'\left(re^{i\theta} + a_j\right)|^2 d\theta + \sup_{0<r<r_k} \int_0^{2\pi} |f_j''\left(re^{i\theta} + a_j\right)|^2 d\theta.$$

Using the above definition we can consider the space $\mathcal{H}^{(2,2)} := \mathcal{H}^{(2,2)}\left(\bigcup_{j=1}^{n} D_k\right)$ as the space of all functions f analytic in $\bigcup_{j=1}^{n} D_k$, endowed with the norm

$$\|f\|^2_{\mathcal{H}^{(2,2)}} := \sum_{j=1}^{n} \|f\|^2_{\mathcal{H}^{(2,2)}(D_j)}$$

3 Statement of the Elastic Problem

We introduce the following constants

$$B_0 = \frac{\sigma_{xx}^{\infty} + \sigma_{yy}^{\infty}}{4}, \text{ and } \Gamma_0 = \frac{\sigma_{yy}^{\infty} - \sigma_{xx}^{\infty} + 2i\sigma_{xy}^{\infty}}{2},$$

where $\sigma_{xx}^{\infty}, \sigma_{xy}^{\infty}, \sigma_{yy}^{\infty}$ are the given stresses at infinity. Consider the functions $\varphi_0(z) = B_0 z + \varphi(z)$, and $\psi_0(z) = \Gamma_0 z + \psi(z)$, where $\varphi(z)$ and $\psi(z)$ are analytical in D and bounded at infinity, $\varphi_k(z)$ and $\psi_k(z)$ are analytical in D_k and all of them are twice differentiable in the closure of the considered domains.

The components of the stress tensor can be determined by the Kolosov-Muskhelishvili formulae [5]

$$\sigma_{xx} + \sigma_{yy} = \begin{cases} 4Re\varphi_k'(z), \ z \in D_k, \\ 4Re\varphi_0'(z), \ \ z \in D, \end{cases} \tag{3.1}$$

$$\sigma_{xx} - \sigma_{yy} + 2i\sigma_{xy} = \begin{cases} -2\left[\overline{z\varphi_k''(z)} + \overline{\psi_k'(z)}\right], \ z \in D_k, \\ -2\left[\overline{z\varphi_0''(z)} + \overline{\psi_0'(z)}\right], \ \ z \in D. \end{cases}$$

The components u, v of the displacement vector are given by

$$u + iv = \begin{cases} \frac{1}{2\mu_k}\left(\kappa_k\varphi_k(z) - \overline{z\varphi_k'(z)} - \overline{\psi_k(z)}\right), \ z \in D_k, \\ \frac{1}{2\mu}\left(\kappa\varphi_0(z) - \overline{z\varphi_0'(z)} - \overline{\psi_0(z)}\right), \ \ z \in D. \end{cases} \tag{3.2}$$

Assuming perfect bonding at the matrix-inclusion interface, the potentials are linked to ∂D_k, $k = 1, ..., n$, through the following conditions

$$\varphi_k(t) + t\overline{\varphi_k'(t)} + \overline{\psi_k(t)} = \varphi_0(t) + t\overline{\varphi_0'(t)} + \overline{\psi_0(t)}, \tag{3.3}$$

$$\mu\left(\kappa_k\varphi_k(t) - t\overline{\varphi_k'(t)} - \overline{\psi_k(t)}\right) = \mu_k\left(\kappa\varphi_0(t) - t\overline{\varphi_0'(t)} - \overline{\psi_0(t)}\right). \tag{3.4}$$

In these equations, μ denotes a shear modulus, κ is the Kolosov constant with $\kappa = 3 - 4\nu$ in plane strain; $\kappa = (3 - \nu)/(1 + \nu)$ in plane stress; ν is Poisson's ratio.

4 Functional-Differential Equations

Introduce the function

$$\Phi_k(z) = \overline{z_{(k)}^{*}}\varphi_k'(z) + \psi_k(z), \ |z - a_k| \leq r,$$

analytic in D_k except point a_k, where its principal part has the form $r^2 (z - a_k)^{-1} \varphi_k'$ (a_k). Following [10] and [18], we can reduce problem (3.3), (3.4) to the system of functional equations

$$\left(1+\frac{\mu}{\mu_k}\kappa_k\right)\varphi_k(z)=\sum_{m\neq k}\left(1-\frac{\mu}{\mu_m}\right)\left[\overline{\Phi_m(z^*_{(m)})}-(z-a_m)\,\overline{\varphi'_m(a_m)}\right]-$$

$$-\left(1-\frac{\mu}{\mu_k}\right)(z-a_k)\,\overline{\varphi'_k(a_k)}+(\kappa+1)B_0 z+p_0,$$

$$|z-a_k|\leq r_k,\ k=1,2,\ldots,n. \tag{4.1}$$

$$\left(\kappa+\frac{\mu}{\mu_k}\right)\Phi_k(z)=\sum_{m\neq k}\left\{\left(\kappa-\frac{\mu}{\mu_m}\kappa_m\right)\overline{\varphi_m(z^*_{(m)})}+\right.$$

$$\left.\left(1-\frac{\mu}{\mu_m}\right)\left(\overline{z^*_{(k)}}-\overline{z^*_{(m)}}\right)\left[\left(\overline{\Phi_m(z^*_{(m)})}\right)'-\overline{\varphi'_m(a_m)}\right]\right\}+$$

$$(1+\kappa)B_0\overline{z^*_{(k)}}+(1+\kappa)\Gamma_0 z+\omega(z),\ |z-a_k|\leq r_k,\ k=1,2,\ldots,n. \tag{4.2}$$

where

$$\omega(z)=\sum_{k=1}^{n}\frac{r^2 q_k}{z-a_k}+q_0, \tag{4.3}$$

q_0 is a constant and

$$q_k=\varphi'_k(a_k)\left(\kappa+\frac{\mu}{\mu_k}-1-\frac{\mu}{\mu_k}\kappa_k\right)-\overline{\varphi'_k(a_k)}\left(1-\frac{\mu}{\mu_k}\right),\ k=1,2,\ldots,n. \tag{4.4}$$

The unknown functions $\varphi_k(z)$ and $\Phi_k(z)$ $(k = 1,2,\ldots,n)$ are related by $2n$ Eqs. (4.1)–(4.2).

One can see that the functional equations do not contain integral operators but they contain compositions of $\varphi_k(z)$ and $\Phi_k(z)$ with inversions. These compositions define compact operators in the Banach space $\mathcal{H}^{(2,2)}\left(\bigcup_{j=1}^{n} D_k\right)$. Hence, the functional Eqs. (4.1)–(4.2) can be effectively solved by use of the symbolic computations. Moreover, the functions $\varphi(z)$ and $\psi(z)$ can be found by the formulae

$$(1+\kappa)\,\varphi(z)=\sum_{m=1}^{N}\left(1-\frac{\mu}{\mu_m}\right)\left[\overline{\Phi_m(z^*_{(m)})}-(z-a_m)\,\overline{\varphi'_k(a_k)}\right]+p_0,\ z\in D. \tag{4.5}$$

$$(1+\kappa)\,\psi(z) = -\sum_{m=1}^{N}\left(1-\frac{\mu}{\mu_m}\right)\overline{z^*_{(m)}}\left[\overline{\left(\Phi_m(z^*_{(m)})\right)'} - \overline{\varphi'_m(a_m)}\right] +$$

$$\sum_{m=1}^{N}\left(\kappa - \frac{\mu}{\mu_m}\kappa_m\right)\overline{\varphi_m(z^*_{(m)})} + \omega(z),\ z \in D. \quad (4.6)$$

Theorem 4.1 ([17]) *For sufficiently small coefficients* μ, μ_k, κ *and* κ_k $(k = 1, \dots, n)$ *the sequence obtained by the method of successive approximations applied to* (4.1)–(4.2) *converges in* $\mathcal{H}^{(2,2)}\left(\bigcup_{j=1}^{n} D_k\right) \times \mathcal{H}^{(2,2)}\left(\bigcup_{j=1}^{n} D_k\right)$.

5 Local Problem

Applying the method of successive approximations we obtain the zeroth order approximation of the form [17]

$$\varphi_k^{(0)}(z) = -\frac{1-\frac{\mu}{\mu_k}}{1+\frac{\mu}{\mu_k}\kappa_k}\frac{(\kappa+1)}{2-\frac{\mu}{\mu_k}+\frac{\mu}{\mu_k}\kappa_k}B_0\,(z-a_k) + (\kappa+1)B_0 z + p_0, \quad (5.1)$$

$$\psi_k^{(0)}(z) = (1+\kappa)\Gamma_0 z + q_0,\ |z-a_k| \le r_k, k = 1, 2, \dots, n. \quad (5.2)$$

Let the approximation of the order $(p-1)$ be known. Then the p-th approximation can be found from a system of differential equations. The equation for $\varphi_k^{(p)}(z)$ has the form

$$\varphi_k^{(p)}(z) = \left(1+\frac{\mu}{\mu_k}\kappa_k\right)^{-1} \times$$

$$\left(\sum_{m\neq k}\left(1-\frac{\mu}{\mu_m}\right)\left[\overline{\Phi_m^{(p-1)}(z^*_{(m)}) - (z-a_m)\left(\varphi_m^{(p-1)}\right)'(a_m)}\right] - \right.$$

$$\left. -\frac{\left(1-\frac{\mu}{\mu_k}\right)(\kappa+1)B_0}{2-\frac{\mu}{\mu_k}+\frac{\mu}{\mu_k}\kappa_k}(z-a_k) + (\kappa+1)B_0 z + p_0\right) \quad (5.3)$$

and the equation for $\psi_k^{(p)}(z)$ has the form

$$\psi_k^{(p)}(z) = \left(\kappa+\frac{\mu}{\mu_k}\right)^{-1}\left(\omega_k^{(p)}(z) + \right.$$

$$\left.\left(1+\frac{\mu}{\mu_k}\kappa_k - \kappa - \frac{\mu}{\mu_k}\right)\overline{z^*_{(k)}}\left(\left(\varphi_k^{(p)}\right)'(z) - \left(\varphi_k^{(p)}\right)'(a_k)\right)\right)$$

$k = 1, 2, \ldots, N$, where $\omega_k^{(p)}$ are calculated by

$$\omega_k^{(p)}(z) = \sum_{m \neq k} \left(\kappa - \frac{\mu}{\mu_m} \kappa_m \right) \overline{\varphi_m^{(p-1)}(z_{(m)}^*)}$$
$$+ \sum_{m \neq k} \left(1 - \frac{\mu}{\mu_m} \right) \overline{z_{(m)}^*} \left[\overline{\left(\Phi_m^{(p-1)}(z_{(m)}^*) \right)'} - \overline{\left(\varphi_m^{(p-1)} \right)' (a_m)} \right] +$$
$$\sum_{m \neq k} \frac{r_m^2}{z - a_m} q_m + (1 + \kappa) \Gamma_0 z + q_0, \; |z - a_k| \leq r_k, \; k = 1, 2, \ldots, N. \tag{5.4}$$

Applying the above results we can present the stress tensor coefficients via potentials. For example the stress tensor coefficient σ_{xx} has the form

$$\sigma_{xx} = \begin{cases} 2Re \left(2\varphi_k' - z\overline{\varphi_k''(z)} - \overline{\psi_k'(z)} \right), \; z \in D_k, \\ 2Re \left(2\varphi_0' - z\overline{\varphi_0''(z)} - \overline{\psi_0'(z)} \right), \; z \in D. \end{cases} \tag{5.5}$$

In the figure Fig. 1 we present contour plot of the above stress tensor component for random distributed inclusions. The inclusions, all of radii 0.1, have the coordinates $\{-0.7399 - 0.08001\imath, -0.2698 - 0.463\imath, 0.4846 - 0.5152\imath, 0.2698 - 0.05099\imath, -0.1712 + 0.2392\imath, 0.5368 + 0.4829\imath, 0.08415 + 0.7498\imath\}$, where we use Euler's notation for the imaginary unit number: $\imath^2 = -1$.

6 Doubly Periodic Problem

Consider the doubly periodic problem, with the periods ω_1, ω_2 whose ratio ω_2/ω_1 is not a real number. It should be mentioned that Jacobi's theorem says that there is no single-valued analytic function with more than two periods. The classical elliptic functions with the above periods are called the Weierstrass $\wp$ and ζ—functions. The Weierstrass $\wp$-function can be represented in the form of the series

$$\wp(z) = \frac{1}{z} + \sum_{m_1, m_2}{}' \left[\frac{1}{(z - m_1\omega_1 - m_2\omega_2)^2} - \frac{1}{(m_1\omega_1 + m_2\omega_2)^2} \right]. \tag{6.1}$$

Following Weil [19] and Akhiezer [20], and using the Eisenstein summation method

$$\sum_{m_1, m_2} := \lim_{M_2 \to \infty} \sum_{m1 = -M_2}^{M_2} \left(\lim_{M_1 \to \infty} \sum_{m_1 = -M_1}^{M_1} \right), \tag{6.2}$$

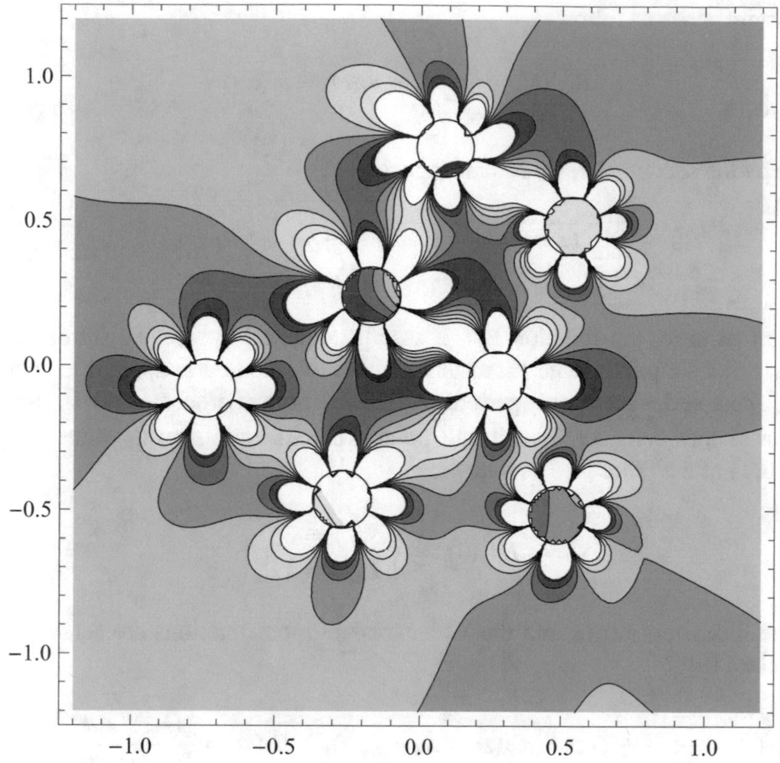

Fig. 1 Contours plot of the σ_x stress tensor component for random distributed inclusions ($\mu = 1.12782, \mu_1 = 33.3333, \kappa = 1.68, \kappa_1 = 2.2$)

we introduce the Eisenstein–Rayleigh lattice sum of the form

$$S_k(\omega_1, \omega_2) = \sum_{m_1,m_2}{}' (m_1\omega_1 + m_2\omega_2)^{-k} \tag{6.3}$$

and the Natanzon–Rayleigh lattice sum

$$T_k(\omega_1, \omega_2) = \sum_{m_1,m_2}{}' \overline{(m_1\omega_1 + m_2\omega_2)}(m_1\omega_1 + m_2\omega_2)^{-k}. \tag{6.4}$$

The symbol $\sum_{m_1,m_2}'$ means that m_1 and m_2 run over all integer numbers, except for the pair $m_1 = m_2 = 0$.

The series above play a pivotal role in the effective approach in the conductivity theory and stress theory [16, 21]. By means of the first sum we can define the

Eisenstein series as follows

$$E_k^{(0)}(z) = \sum_{m_1,m_2} (z - m_1\omega_1 - m_2\omega_2)^{-k} \tag{6.5}$$

while by the second one, the Natanzon series

$$E_k^{(1)}(z) = \sum_{m_1,m_2} \overline{(z - m_1\omega_1 - m_2\omega_2)}\,(z - m_1\omega_1 - m_2\omega_2)^{-k}. \tag{6.6}$$

In the literature, the function $E_k^{(0)}$ is usually denoted by E_k. We introduce new notation for simplicity. The Eisenstein summation method is applied to (6.5) and (6.6). These series are doubly periodic and have a pole of order k at $z = 0$. However, for our further considerations it will be convenient to define the value of $E_k^{(j)}(z)$ ($j = 0, 1$) at the point zero as follows:

$$E_k^{(j)}(0) := \begin{cases} S_k, \, j = 0, \\ T_k, \, j = 1. \end{cases} \tag{6.7}$$

The Eisenstein series and the Weierstrass elliptic functions are related by the formulae [19]

$$E_1(z) = -zS_2 + \zeta(z), \quad E_2(z) = S_2 + \wp(z), \quad E_n(z) = \frac{(-1)^n}{(n-1)!}\frac{d^{n-2}}{dz^{n-2}}\wp(z)$$

The Natanzons series are related with the Weiestrass functions by the following formula obtained in [16]

$$E_3^{(1)}(z) = -\frac{1}{2}\bar{z}\wp'(z) + \frac{1}{6\pi}\wp''(z) + \frac{1}{2}\left[\frac{\zeta(z)}{\pi} - \left(\frac{S_2}{\pi} - 1\right)z\right]\wp'(z)$$
$$-\left(\frac{S_2}{\pi} - 1\right)\wp(z) - \frac{5}{\pi}S_4 + S_3^{(1)}. \tag{6.8}$$

Let l be a natural number; k_l runs over 1 to N, $m_l = 2, 3, \ldots, l = 0, .., q$. Let $\mathbf{C}$ be the operator of complex conjugation. The generalized Eisenstein–Rayleigh sum (e-sum) is defined as follows [22, 23]

$$e_{m_1 \ldots m_l}^{(j_1 .. j_q)} = \frac{1}{N}(N\pi)^{-\left[1+\frac{1}{2}\left((m_1-j_1)+\cdots+(m_q-j_q)\right)\right]} \times$$
$$\sum_{k_0,k_1,\ldots,k_q} E_{m_1}^{(j_1)}(a_{k_0} - a_{k_1})\,\overline{E_{m_2}^{(j_2)}(a_{k_1} - a_{k_2})\ldots \mathbf{C}^{q+1}\, E_{m_q}^{(j_q)}}\left(a_{k_{q-1}} - a_{k_q}\right). \tag{6.9}$$

We write $e_{m_1 \ldots m_q}^{(0 \ldots 0)} = e_{m_1 \ldots m_q}$, and $e_{m_1 \ldots m_q}^{(1 \ldots 1)} = e_{m_1 \ldots m_q}^{(1)}$ for short.

The number of sums in formula (6.9) is equal to $(q+1)$ and can be reduced in the following way (for a similar method see [23]). Define the function

$$F_p^{(1)}(z) = \frac{1}{N}\sum_{k=1}^{N} E_p^{(1)}(z - a_k).$$

Note that

$$E_p^{(1)}(-z) = (-1)^{p+1}E_p^{(1)}(z). \tag{6.10}$$

Then the sum $e_{pp}^{(1)}$ can be written in the form

$$e_{pp}^{(1)} = \frac{1}{(N\pi)^p}\sum_{k_0k_1} E_p^{(1)}(a_{k_0} - a_{k_1})\overline{F_p^{(1)}(a_{k_1})} =$$

$$\frac{(-1)^{p+1}}{N^{p-1}\pi^p}\sum_{k_1} F_p^{(1)}(a_{k_1})\overline{F_p^{(1)}(a_{k_1})} = \frac{(-1)^{p+1}}{N^{p+1}\pi^p}\sum_{k_1}\left|\sum_k E_p^{(1)}(a_{k_1} - a_k)\right|^2 \tag{6.11}$$

One can see that $e_{pp} \leq 0$ for even p and $e_{pp} \geq 0$ for odd p.

Lemma 6.1 *Let j be a natural number equal to* 0 *or* 1, $\alpha = \sum_{s=1}^{q} m_s - qj$. *Then*

$$e_{m_1\ldots m_q}^{(j)} = (-1)^{\alpha}\mathbf{C}^{q+1}e_{m_q\ldots m_1}^{(j)}. \tag{6.12}$$

Remark 6.2 This lemma was proved in [24] for $s = 0$.

Proof Let $j = 1$. From (6.9) we get

$$e_{m_1\ldots m_q}^{(1)} = \frac{1}{N}(N\pi)^{-(1+\frac{\alpha}{2})}\sum_{k_0,k_1,\ldots,k_q} E_{m_1}^{(j)}(-(a_{k_1} - a_{k_0}))\times$$

$$\overline{E_{m_2}^{(j)}(-(a_{k_2} - a_{k_1}))}\ldots\mathbf{C}^{q+1}E_{m_q}^{(j)}\left(-(a_{k_q} - a_{k_{q-1}})\right). \tag{6.13}$$

It follows from (6.10) that the right-hand side expression of (6.13) is of the form

$$(-1)^{\alpha}\frac{1}{N}(N\pi)^{-(1+\frac{\alpha}{2})}\sum_{k_0,k_1,\ldots,k_q}\mathbf{C}^{q+1}E_{m_q}^{(j)}\left(a_{k_q} - a_{k_{q-1}}\right)\ldots\overline{E_{m_2}^{(j)}(a_{k_2} - a_{k_1})}E_{m_1}^{(j)}(a_{k_1} - a_{k_0})$$

which is equal to $(-1)^{\alpha}e_{m_q\ldots m_1}^{(1)}$ or $(-1)^{\alpha}\overline{e_{m_q\ldots m_1}^{(1)}}$, since q is odd or even respectively. The lemma is proved. □

The Eisenstein and Natanzons series are defined for the fixed fundamental vectors ω_1, ω_2. A change of the scale is equivalent to the passage from a lattice to a sublattice. More precisely, let $\mathcal{G}$ be a sublattice of $\mathcal{G}'$. Then the fundamental vectors ω_1, ω_2 of $\mathcal{G}$, and ω_1', ω_2' of $\mathcal{G}'$ are related by the equation

$$(\omega_1, \omega_2) = (\omega_1', \omega_2') \cdot A \tag{6.14}$$

where the matrix A consists of the integer components [19] and $detA = N$. The factor $\mathcal{G}/\mathcal{G}'$ consists of N points $a_1, a_2, \ldots, a_N$ lying in the cell $\mathcal{G}$. These points also belong to $\mathcal{G}'$, that is, each a_j is a linear combination of ω_1', ω_2' with the integer coefficients. Let $E_p^{(1)}(z; \omega_1, \omega_2)$ be the generalized Eisenstein function of order p associated with the periods ω_1 and ω_2. Then in view of [19, 23] we obtain

$$\sum_{w \in \mathcal{G}/\mathcal{G}'} E_p^{(1)}(z + w; \omega_1, \omega_2) = E_p^{(1)}(z; \omega_1', \omega_2').$$

In particular, we have

$$\sum_{w \in \mathcal{G}/\mathcal{G}'} E_p^{(1)}(w; \omega_1, \omega_2) = S_p^{(1)}(\omega_1', \omega_2').$$

Without loss of generality we can assume that the area $|\mathcal{G}|$ of the fundamental cell $\mathcal{G}$ is equal to 1 and $|\mathcal{G}'| = N^{-1}$. Then the change of the linear scale in (6.3) yields

$$S_p^{(1)}(\omega_1', \omega_2') = N^{\frac{p-1}{2}} S_p^{(1)}(\omega_1, \omega_2).$$

In [23], a result was presented for the standard Eisenstein sum of the form

$$S_p(\omega_1', \omega_2') = N^{\frac{p}{2}} S_p(\omega_1, \omega_2).$$

The both results yield

$$S_p^{(j)}(\omega_1', \omega_2') = N^{\frac{p-j}{2}} S_p^{(j)}(\omega_1, \omega_2),$$

where $j = 0, 1$.

7 Effective Formula

Consider now two phase material. For any effective formula the exact Hashin–Strikman restriction is known

$$\mu^- \leq \mu_e \leq \mu^+, \quad k^- \leq k_e \leq k^+, \tag{7.1}$$

where

$$k^{-} = k + \frac{f}{\frac{1}{k_1-k} + \frac{1-f}{k+\mu}}, \quad k^{+} = k_1 + \frac{1-f}{\frac{1}{k-k_1} + \frac{f}{k_1+\mu_1}}, \tag{7.2}$$

$$\mu^{-} = \mu + \frac{f}{\frac{1}{\mu_1-\mu} + \frac{(1-f)(k+2\mu)}{2\mu(k+\mu)}}, \quad \mu^{+} = \mu_1 + \frac{1-f}{\frac{1}{\mu-\mu_1} + \frac{f(k_1+2\mu_1)}{2\mu_1(k_1+\mu_1)}}. \tag{7.3}$$

The stress and deformation tensors can be calculated by the Kolosov-Muskhelishvili formulae [5]. Then, the effective elastic moduli of macroscopically isotropic fibrous composites can be calculated as follows

$$\mu_{eff} = \frac{\langle\sigma_{xx} - \sigma_{yy}\rangle}{2\langle\varepsilon_{xx} - \varepsilon_{yy}\rangle}, \; k_{eff} = \frac{\langle\sigma_{xx} + \sigma_{yy}\rangle}{2\langle\varepsilon_{xx} + \varepsilon_{yy}\rangle}. \tag{7.4}$$

Here, the limit average over the plane is introduced

$$\langle\cdot\rangle = \lim_{Q_n\to\infty} \frac{1}{|Q_n|} \iint\limits_{Q_n} \cdot dxdy.$$

In the latter limit, it is assumed that the infinitely many points a_k are distributed in the plane. After long symbolic computations we get

$$\begin{aligned}\mu_{eff} = \mu - &\frac{(\kappa+1)\mu\,(\mu-\mu_1)}{\kappa\mu_1+\mu}f + \\ &\left(-\frac{2B_0(\kappa+1)\mu(\mu-\mu_1)(-\kappa\mu_1+(\kappa_1-1)\mu+\mu_1)}{\pi\Gamma_0(\kappa\mu_1+\mu)((\kappa_1-1)\mu+2\mu_1)} + \right. \\ &\left.\frac{2e_3^{(1)}(\kappa+1)\mu\overline{\Gamma_0}(\mu-\mu_1)^2}{\pi\Gamma_0(\kappa\mu_1+\mu)^2} + \frac{\kappa(\kappa+1)\mu(\mu-\mu_1)^2}{(\kappa\mu_1+\mu)^2}\right) f^2 + O(f^3)\end{aligned} \tag{7.5}$$

and

$$\begin{aligned}k_{eff} = k\Bigg(1 + &\frac{(\kappa+1)\mu\,(-\kappa_1\mu+(\kappa-1)\mu_1+\mu)}{(\kappa-1)\,((\kappa_1-1)\,\mu+2\mu_1)}f \\ &+\left(\frac{e_2^{(0)}(\kappa+1)\,(\mu-\mu_1)\,\overline{\Gamma_0}\,(\kappa_1\mu+\mu_1)\,(-\kappa_1\mu+\kappa\mu_1+\mu-\mu_1)}{\pi B_0(\kappa-1)\,(\kappa_1+1)\,(\kappa_1\mu-\mu+2\mu_1)\,(\kappa\mu_1+\mu)}\right. \\ &+\frac{\overline{e_2^{(0)}}\Gamma_0(\kappa+1)\,(\mu-\mu_1)^2\,(-\kappa_1\mu+(\kappa-1)\mu_1+\mu)}{\pi B_0(\kappa-1)\,(\kappa_1+1)\,((\kappa_1-1)\,\mu+2\mu_1)\,(\kappa\mu_1+\mu)} \\ &\left.-\frac{2(\kappa+1)\mu\,(-\kappa_1\mu+(\kappa-1)\mu_1+\mu)\,((\kappa_1-1)\,\mu-\kappa\mu_1+\mu_1)}{(\kappa-1)^2\,((\kappa_1-1)\,\mu+2\mu_1)^2}\right)f^2 + O(f^3)\Bigg).\end{aligned} \tag{7.6}$$

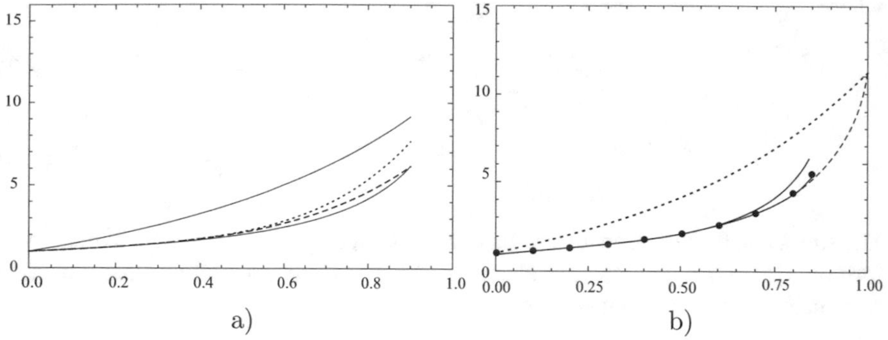

Fig. 2 The dependencies of $\frac{k_e}{k}$ on the concentration f for the regular hexagonal lattice composites. (**a**) Dashed lines correspond to accuracy $O(f^6)$; dotted line corresponds to accuracy $O(f^7)$; solid line corresponds to the Hashin–Shtrikman bounds (7.2), (7.6) ($\mu = 1, \mu_1 = 22.5, k = 3.33, k_1 = 37.5$), (**b**) shows Fig. 5 from [25], where dashed line for two-point bound, solid line for three-point bound and the black circles for BEM results

Note that the coefficients of the bulk and shear moduli depend on B_0 and Γ_0. The homogenization theory implies that it is impossible. In the paper [16], it is proved that $\frac{B_0}{\Gamma_0}S_2$ and $\frac{\overline{\Gamma_0}}{\Gamma_0}S_3^{(1)}$ are invariant under the change of B_0 and Γ_0. This means that the dependence of the effective coefficients on B_0 and Γ_0 is illusory.

Consider fibrous regular composite material composed of resin with mechanical properties $\mu = 1$, $\mu_1 = 22.5$, $k = 3.33$, $k_1 = 37.5$ (as in [25]). The effective bulk moduli for this data is calculated by formula

$$k_e = 3.33 + 3.84302f + 3.4108f^2 + 4.21712f^3 + 5.49012f^4 + 7.19337f^5 + 9.43231f^6.$$

In Fig. 2 we present dependencies of $\frac{k_e}{k}$ on the concentration f for the regular hexagonal lattice composites.We observe improvement of the approximation for larger value of the concentration. Data coincides with data in [25]. The results of the two-point method (Fig. 2b) coincides with Hashin–Shtrikman bounds presented in the Fig. 2a.

In the paper [16], different, random composites are presented. Other interesting result are presented in papers [26–34].

Acknowledgements The research has been partially supported by the Centre for Innovation and Transfer of Natural Science and Engineering Knowledge of University of Rzeszów (grant No. WMP/GD-09/2016).

References

1. J. Byström, N. Jekabsons, J. Varna, An evaluation of different models for prediction of elastic properties of woven composites. Compos. Part B **31**, 7–20 (2000)
2. S. Berggren, D. Lukkassen, A. Meidell, L. Simula, Some methods for calculating stiffness properties of periodic structure. Appl. Math. **48**(2), 97–110 (2003)
3. J. Franců, Homogenization of linear elasticity equations. Apl. Mat. **27**, 96–117 (1982)
4. L. Greengard, J. Helsing, On the numerical evaluation of elastostatic fields in locally isotropic two-dimensional composites. J. Mech. Phys. Solids **46**, 1441–1462 (1998)
5. N.I. Muskhelishvili, *Some Basic Problems of the Mathematical Theory of Elasticity*, Reprint of the 2nd English edn. (Springer-Science + Business Media, Dordrecht, 1977)
6. I. Jasiuk, J. Chen, M.F. Thorpe, *Elastic Properties of Two-Dimensional Composites Containing Polygonal Holes*. Materials Division, vol. 35 (American Society of Mechanical Engineers, New York, 1992), pp. 61–73
7. A.M. Linkov, *Boundary Integral Equations in Elasticity Theory* (Kluwer Academic Publishers, Dordrecht, 2002)
8. J. Wang, S.L. Crouch, S.G. Mogilevskaya, A complex boundary integral method for multiple circular holes in an infinite plane. Eng. Anal. Bound. Elem. **27**(8), 789–802 (2003)
9. S.G. Mogilevskaya, V.I. Kushch, H.K. Stolarski, S.L. Crouch, Evaluation of the effective elastic moduli of tetragonal fiber-reinforced composites based on Maxwell's concept of equivalent inhomogeneity. Int. J. Solids Struct. **50**(25–26), 4161–4172 (2013)
10. V. Mityushev, S.V. Rogosin, *Constructive Methods for Linear and Nonlinear Boundary Value Problems for Analytic Functions. Theory and Applications* (Chapman and Hall/CRC, London, 1999)
11. P. Drygaś, Generalized Eisenstein functions. J. Math. Anal. Appl. **444**(2), 1321–1331 (2016)
12. V.Ya. Natanson, On the stresses in a stretched plate weakened by identical holes located in chessboard arrangement. Mat. Sb. **42**(5), 616–636 (1935)
13. L.A. Filshtinsky, V. Mityushev, *Mathematical Models of Elastic and Piezoelectric Fields in Two-Dimensional Composites*, ed. by P.M. Pardalos, T.M. Rassias. Mathematics Without Boundaries (Springer, New York, 2014), pp. 217–262
14. E.I. Grigolyuk, L.A. Filshtinsky, *Perforated Plates and Shells* (Nauka, Moscow, 1970); [in Russian]
15. E.I. Grigolyuk, L.A. Filshtinsky, *Periodic Piecewise-Homogeneous Elastic Structures* (Nauka, Moscow, 1992); [in Russian]
16. P. Drygaś, V. Mityushev, Effective elastic properties of random two-dimensional composites. Int. J. Solids Struct. **97–98**, 543–553 (2016)
17. P. Drygaś, Functional-differential equations in a class of analytic functions and its application to elastic composites. Complex Variables Elliptic Equ. **61**(8), 1145–1156 (2016)
18. V. Mityushev, Thermoelastic plane problem for material with circular inclusions. Arch. Mech. **52**(6), 915–932 (2000)
19. A. Weil, *Elliptic Functions According to Eisenstein and Kronecker* (Springer, Berlin, 1976)
20. N.I. Akhiezer, *Elements of the Theory of Elliptic Functions* (American Mathematical Society, Providence, RI, 1990)
21. V.V. Mityushev, E. Pesetskaya, S.V. Rogosin, *Analytical Mathods for Heat Conduction in Composites and Porous Media*, ed. by A. Ochsner, G.E. Murch, M.J.S. de Lemos (Wiley, New York, 2008)
22. V.V. Mityushev, Representative cell in mechanics of composites and generalized Eisenstein's-Rayleigh sums. Complex Variables **51**(8–11), 1033–1045 (2006)
23. V. Mityushev, N. Rylko, Optimal distribution of the nonoverlapping conducting disks. Multiscale Model. Simul. **10**(1), 180–190 (2012)
24. R. Czapla, W. Nawalaniec, V. Mityushev, Effective conductivity of random two-dimensional composites with circular non-overlapping inclusions. Comput. Mater. Sci. **63**, 118–126 (2012)

25. J.W. Eischen, S. Torquato, Determining elastic behavior of composites by the boundary element method. J. Appl. Phys. **74**, 159–170 (1993)
26. P. Drygaś, V. Mityushev, Effective conductivity of unidirectional cylinders with interfacial resistance. Q. J. Mech. Appl. Math. **62**, 235–262 (2009)
27. S. Yakubovich, P. Drygaś, V. Mityushev, Closed-form evaluation of 2D static lattice sums. Proc. R. Soc. A **472**, 20160510 (2016); https://doi.org/10.1098/rspa.2016.0510
28. P. Drygaś, Steady heat conduction of material with coated inclusion in the case of imperfect contact. Math. Model. Anal. **12**(3), 291–296 (2007)
29. P. Drygaś, A functional-differential equation in a class of analytic functions and its application. Aequationes Math. **73**(3), 22–232 (2007)
30. P. Drygaś, Functional-differential equations in Hardy-type classes. Tr. Inst. Mat. **15**(1), 105–110 (2007)
31. V.V. Mityushev, Exact solution of the R-linear problem for a disc in a class of doubly periodic functions. J. Appl. Funct. Anal. **2**(2), 115–127 (2007)
32. V. Mityushev, Transport properties of two-dimensional composite materials with circular inclusions. Proc. R. Soc. Lond. A **455**, 2513–2528 (1999)
33. V.V. Mityushev, E. Pesetskaya, S.V. Rogosin, Analytical methods for heat conduction in composites and porous media, in *Cellular and Porous Materials: Thermal Properties Simulation and Prediction*, ed. by A. Öchsner, G.E. Murch, M.J.S. de Lemos (Wiley, Weinheim, 2008)
34. V. Mityushev, N. Rylko, Maxwell's approach to effective conductivity and its limitations. Q. J. Mech. Appl. Math. **66**(2), 241–251 (2013)

Statistical Characteristics of the Distraction Parameters in the Unbounded Anisotropic Plane Weakened by Multiple Random Cracks

L.A. Filshtinskii, D.M. Nosov, and Yu.V. Shramko

Abstract The boundary value problem of the theory of elasticity for a finite anisotropic plate with random cracks has been solved. Stress intensity factors and energy flows near the tips of cracks are determined as a linear functional on solutions to a system of singular integral equations. It is shown, that in case of the normal distribution of the cracks, the statistical characteristics (mathematical expectations and dispersions) of the distraction (stress intensity factors and energy flows) have also the normal law distribution.

Keywords Anisotropy • Multiple cracks • Singular integral equations • Stress intensity factors

Mathematics Subject Classification (2010) Primary 74R74; Secondary 74B74

1 Introduction

In this paper, we use a numerical approach to solve the boundary value problem of the fracture mechanics for the multiple cracks located in a finite anisotropic plate. Such problems with cracks in homogeneous and composite anisotropic media which were solved by the method of the singular integral equations and Green's function due to Grygolyuk and Filshtinskii [1]. The solution to the problem with

L.A. Filshtinskii • Yu.V. Shramko
Rymskogo-Korsacova. 2, Sumy State University, 40007 Sumy, Ukraine
e-mail: leonid@mphis.sumdu.edu.ua; yshramko@pom.sumdu.edu.ua

D.M. Nosov (✉)
Pedagogical University of Cracow, Podchorazych. 2, 30-084 Cracow, Poland
e-mail: d.nosov@pom.sumdu.edu.ua; bate8075@gmail.com

P. Drygaś, S. Rogosin (eds.), *Modern Problems in Applied Analysis*,
Trends in Mathematics, https://doi.org/10.1007/978-3-319-72640-3_8

cracks in anisotropic media was obtained by using the Stroh formalism and the boundary elements method, namely, the generalized plane strain problem by Denda [2] and for the cracks between the dissimilar anisotropic materials by Ikeda et al. [3]. The simulation of the local fields near the cracks and inclusions in strongly inhomogeneous materials were performed by the boundary element method due to Rejwer et al. [4]. The Galerkin boundary integral method for the combinations of cracks in the structure for the computation stress intensity factors and fields in the infinite anisotropic plane was used in [5, 6]. The numerical experiment consists in the repeated computations of random normally distributed cracks to create the statistical sample which contains the statistical characteristics of distraction (mathematical expectations, dispersions). In the present paper, the distributions of the parameters of distraction are obtained by the application of the Monte Carlo scheme.

2 Problem Statement

We consider the elastic anisotropic plane weakened in some finite area by the multiple cracks $\Gamma_n, n = \overline{1, M}$ in Cartesian axes x_1, x_2. We assume that the cracks are located randomly. Every crack Γ_n is the Lyapunov's arc and $\bigcap \Gamma_n = \emptyset$. Let us suppose that the normal pressure p_n is loaded on the sides of Γ_n. The uniform field of the normal and tangential stresses σ_{ij}^{∞} is given at infinity (Fig. 1).

The purpose of the present research is the development of the effective analytic-numerical algorithm to investigate the deformation-stress state at every point of the body and identify SIF (stress intensity factor) and the energy flows at the tips of the cracks at the numerical experiment conditions.

The short review of the plane stressed state of the anisotropic media is given in [7].

The Hooke's law reads as follows:

$$\begin{gathered} e_{11} = s_{11}\sigma_{11} + s_{12}\sigma_{22} + s_{16}\sigma_{12} \\ e_{22} = s_{12}\sigma_{11} + s_{22}\sigma_{22} + s_{26}\sigma_{12} \\ 2e_{12} = s_{16}\sigma_{11} + s_{26}\sigma_{22} + s_{66}\sigma_{12} \\ e_{ij} = (\partial_i u_j + \partial_j u_i)/2; \quad (i, j = 1, 2, 3) \\ \partial_i = \partial/\partial x_i \quad z_k = x_1 + \mu_k x_2 \end{gathered} \tag{2.1}$$

where s_{ij} are the deformation coefficient, $u_j, (j = 1, 2)$– the components of the elastic dislocations vector, σ_{ij}, e_{ij} the components of the stress and deformation

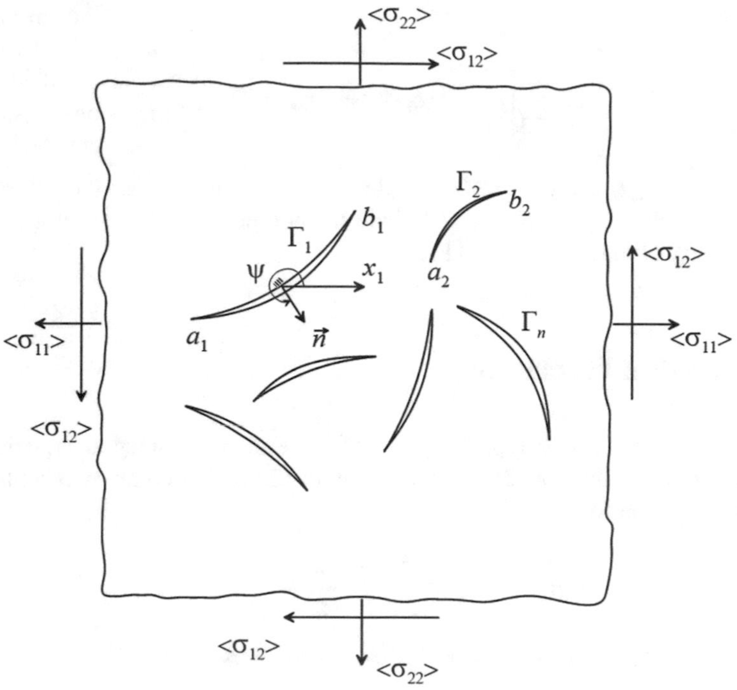

Fig. 1 Cracks and loads in anisotropic plane

tensors, respectively,

$$\{\sigma_{11}, \sigma_{12}, \sigma_{22}\} = 2\Re \sum_{k=1}^{2} \{\mu_k^2, -\mu_k, 1\}\Phi_k(z_k);$$

$$\{u_1, u_2\} = 2\Re \sum_{k=1}^{2} \{p_{1k}, p_{2k}\}\varphi_k(z_k);\;\; \Phi_k(z_k) = \frac{d}{dz_k}\varphi_k(z_k); \tag{2.2}$$

$$p_{1k} = s_{11}\mu_k^2 + s_{12} - s_{16}\mu_k; p_{2k} = s_{12}\mu_k + s_{22}/\mu_k - s_{26};$$

The numbers μ_k here are the simple roots of the algebraic equation of the fourth degree. It is assumed that the roots are different in the upper half-plane.

$$s_{11}\mu^4 - 2s_{16}\mu^3 - (2s_{12} - s_{16})\mu^2 - 2s_{26}\mu + s_{22} = 0;$$
$$\Im\mu_1 > 0, \Im\mu_2 > 0, \mu_3 = \overline{\mu_1}, \mu_4 = \overline{\mu_2} \tag{2.3}$$

The bar means the complex conjugation. Thereby, the dislocations, stresses and deformations in the body are expressed through two functions $\varphi_k(z_k)$, which are analytical in the corresponding regions. The boundary conditions can be written in the form

$$2\Re\sum_{k=1}^{2}\{\mu_k,-1\}a_k(\psi)\Phi_k(z_k)=\{X_{1n},X_{2n}\};\tag{2.4}$$

where $a_k(\psi)=\mu_k cos(\psi)-sin(\psi)$, ψ denotes the angle between the outer normal vector $\vec{n}$ and the Ox_1 axis; X_{1n} and X_{2n} are the components of the vector acting on the boundary of the body $\Gamma=\bigcup\Gamma_n$.

3 Analytical Formalism

The generalized Cauchy type integrals and the singular integral equations [8] are used to solve the boundary value problem (2.4). The complex potentials are represented in the form

$$\begin{gathered}\Phi_k(z_k)=A_k+\frac{1}{2\pi}\int_\Gamma\frac{\omega_k(\zeta)}{(\zeta_k-z_k)}ds\\ z=x_1+ix_2, z_k=\Re z+\mu_k\Im z, \zeta_k=\Re\zeta+\mu_k\Im\zeta, \zeta\in\Gamma\end{gathered}\tag{3.1}$$

Here, ds means the element of the arc in the physical plane z, z_k and ζ_k are the images of the complex variables z and ζ correspondingly, constants A_k have to be found through the uniform field σ_{ij} at infinity.

Taking into account the equalities $d\zeta_k=a_k ds$ we transform the representation (3.1)

$$\Phi_k(z_k)=A_k+\frac{1}{2\pi}\int_\Gamma\frac{\omega_k(\zeta)}{a_k(\psi)(\zeta_k-z_k)}d\zeta_k\tag{3.2}$$

We find the traces of the function $\Phi_k(z_k)$ on Γ by means of the Sokhotskii-Plemelj formulas [9]. The Eq. (3.2) yields

$$\begin{gathered}\{\Phi_k(z_k)\}^{\pm}_{z\to\zeta_0\in\Gamma}=\pm\frac{i\omega_k(\zeta_0)}{2a_k(\psi_0)}+\frac{1}{2\pi}\int_\Gamma\frac{\omega_k(\zeta)}{\zeta_k-\zeta_{0k}}ds+A_k;\\ \zeta_{0k}=\Re\zeta_0+\mu_k\Im\zeta_0\end{gathered}\tag{3.3}$$

The upper sign corresponds to the case $z\leftarrow\zeta_0\in\Gamma=\bigcup\Gamma_n$. The integral is considered as the principal value by Cauchy. Substitution of the limit values (3.3) into the boundary conditions (2.4) yields

$$2\Re\sum_{k=1}^{2}\{\mu_k,-1\}a_k(\psi_0)\{\pm\frac{i\omega_k(\zeta_0)}{2a_k(\psi_0)}+\frac{1}{2\pi}\int_\Gamma\frac{\omega_k(\zeta)}{\zeta_k-\zeta_{0k}}ds\}=\{X_{1n},X_{2n}\}\tag{3.4}$$

Hence, the system consists of two algebraic and two singular integral equations of the first kind with Cauchy kernel.

$$\Im \sum_{k=1}^{2} -\omega_k(\zeta) = 0; \ \Im \sum_{k=1}^{2} \mu_k \omega_k(\zeta) = 0; \tag{3.5}$$

$$2\Re \sum_{k=1}^{2} \{\mu_k, -1\} \frac{a_k(\psi_0)}{\pi} \int_{\Gamma} \frac{\omega_k(\zeta)}{\zeta_k - \zeta_{0k}} ds = \{N_1(\zeta_0), N_2(\zeta_0)\} \tag{3.6}$$

where

$$\begin{gathered} N_1(\zeta_0) = -p\cos(\psi_0) - 2\Re \sum_{k=1}^{2} \mu_k a_k(\psi_0) A_k; \\ N_2(\zeta_0) = -p_n \sin(\psi_0) + 2\Re \sum_{k=1}^{2} a_k(\psi_0) A_k; \ p = \{p_n, \zeta \in \Gamma_n\} \end{gathered} \tag{3.7}$$

It follows from the relations (2.2) and (2.3), that at "infinity" must satisfy the equalities

$$2\Re \sum_{k=1}^{2} A_k = \sigma_{22}^{\infty}; \ 2\Re \sum_{k=1}^{2} \mu_k A_k = -\sigma_{12}^{\infty}; \ 2\Re \sum_{k=1}^{2} \mu_k^2 A_k = \sigma_{11}^{\infty}; \tag{3.8}$$

Hence

$$\begin{aligned} 2\Re \sum_{k=1}^{2} \mu_k A_k &= \sigma_{11}^{\infty} \cos\psi_0 + \sigma_{12}^{\infty} \sin\psi_0; \\ 2\Re \sum_{k=1}^{2} -A_k &= \sigma_{12}^{\infty} \cos\psi_0 + \sigma_{22}^{\infty} \sin\psi_0; \end{aligned} \tag{3.9}$$

We introduce the real vector-function $q(\zeta)$ by the equations

$$\begin{gathered} q(\zeta) = R\omega(\zeta), \ \omega(\zeta) = R^{-1}q(\zeta) \\ \omega(\zeta) = \{\omega_1, \omega_2\}^T, \ q(\zeta) = \{q_1, q_2\}^T, R = \begin{pmatrix} \mu_1 & \mu_2 \\ -1 & -1 \end{pmatrix} \end{gathered} \tag{3.10}$$

Inversion of the system (3.7) gives

$$\omega(\zeta) = (-1)^k \frac{q_1 + \varepsilon_k q_2}{\mu_2 - \mu_1}, \ \varepsilon_1 = \mu_2, \ \varepsilon_2 = \mu_1, \tag{3.11}$$

Ultimately, we reduce the integral Eq. (3.5) using the relations (3.6), (3.7) to the following matrix form

$$\frac{1}{\pi}\int_{\Gamma} K(\zeta,\zeta_0)q(\zeta)ds = N(\zeta_0),\ \zeta_0 \in \Gamma = \bigcup_{1}^{M}\Gamma_n,$$

$$K(\zeta,\zeta_0) = \Re\left[R(G(\zeta,\zeta_0))R^{-1}\right]$$

$$G(\zeta,\zeta_0) = diag\{\frac{a_1(\psi_0)}{\zeta_1-\zeta_{01}},\frac{a_2(\psi_0)}{\zeta_2-\zeta_{02}},\frac{a_3(\psi_0)}{\zeta_3-\zeta_{03}}\},\ a_k(\psi_0) = \mu_k\cos\psi_0 - \sin\psi_0$$

$$q(\zeta) = (q_1(\zeta),q_2(\zeta))^T,\ q_k(\zeta) = \{q_k^n(\zeta),\ \zeta\in\Gamma_n\},\ (k=1,2;\ n=\overline{1,M}) \tag{3.12}$$

The solution to the matrix integral Eq. (3.9) yields the additional conditions of the single-valuedness of dislocations. This conditions can be obtained from (3.1)

$$\varphi_k(z_k) = A_k z_k - \frac{1}{2\pi}\int_{\Gamma}\omega_k(\zeta)\log(\zeta_k - z_k)ds,\ \varphi_k'(z) = \Phi(z) \tag{3.13}$$

The increment of the function $\varphi_k(z_k)$ for the full path tracing along the crack Γ_n follows from (3.10)

$$\Delta\varphi_k(z_k) = -i\int_{\Gamma_n}\omega_k(\zeta)ds \tag{3.14}$$

Using the formulas (2.2) and taking into account (3.11) we obtain the conditions of the single valuedness of the field of displacements in the plate

$$\Im\sum_{k=1}^{2}p_{1k}\int_{\Gamma_n}\omega_k(\zeta)ds = 0;\ \Im\sum_{k=1}^{2}p_{2k}\int_{\Gamma_n}\omega_k(\zeta)ds = 0;\ (n=\overline{1,M}) \tag{3.15}$$

We have

$$\Im\sum_{k=1}^{2}\frac{(-1)^{k-1}p_{1k}}{\varepsilon_2-\varepsilon_1} = s_{11}\Im(\mu_1+\mu_2),\ \Im\sum_{k=1}^{2}\frac{(-1)^{k-1}\varepsilon_k p_{1k}}{\varepsilon_2-\varepsilon_1} = s_{11}\Im(\mu_1\mu_2)$$

$$\Im\sum_{k=1}^{2}\frac{(-1)^{k-1}p_{2k}}{\varepsilon_2-\varepsilon_1} = -s_{22}\Im\left(\frac{1}{\mu_1\mu_2}\right),\ \Im\sum_{k=1}^{2}\frac{(-1)^{k-1}\varepsilon_k p_{2k}}{\varepsilon_2-\varepsilon_1} = -s_{22}\Im\left(\frac{\mu_1+\mu_2}{\mu_1\mu_2}\right) \tag{3.16}$$

Taking into account this relations and (3.9) we reduce the single-valuedness conditions (3.12) to the following ones

$$s_{11}\{\Im(\mu_1+\mu_2)\int_{\Gamma} q_1(\zeta)ds - \Im(\mu_1\mu_2)\int_{\Gamma_n} q_2(\zeta)ds\} = 0; \quad (n=\overline{1,\ M})$$
$$s_{22}\{-\Im\left(\frac{1}{\mu_1+\mu_2}\right)\int_{\Gamma} q_1(\zeta)ds + \Im\left(\frac{\mu_1+\mu_2}{\mu_1\mu_2}\right)\int_{\Gamma_n} q_2(\zeta)ds\} = 0; \tag{3.17}$$

Moreover, $s_{22} = s_{11}|\mu_1\mu_2|^2$.

We can show that the determinant of this homogeneous system is not equal to zero and

$$\int_{\Gamma} q_k(\zeta)ds = 0, \quad (k=1,\ 2;\ n=\overline{1,\ M}) \tag{3.18}$$

Thereby, the constructed analytical algorithm is reduced to the solution to the system of $2M$ matrix singular integral equations of the first kind (3.9) simultaneously with $2M$ additional conditions (3.14). These conditions yield the single-valuedness of the solution to the system (3.9) in the class of the functions unbounded at the ends of the arcs Γ_n, $(n=\overline{1,\ M})$ [9].

4 Asymptotes at the Ends of Cracks and Characteristics of Distraction

Let us introducc the contour parametrization Γ_n : $\zeta = \zeta(\beta)$, $\zeta_0 = \zeta_0(\beta)$, $-1 \le \beta, \beta_0 \ge 1$. The densities ω_k in the Eq. (3.4) have the square root singularities. Let us suppose that

$$\omega_k(\zeta) = \frac{\omega_k^*(\zeta)}{\sqrt{(\zeta-a)(\zeta-b)}} = \frac{\Omega_k(\beta)}{\sqrt{1-\beta^2 s'(\beta)}}, s'(\beta) = \frac{ds}{d\beta}, \ (k=1,2) \tag{4.1}$$

where $\Omega_k(\beta)$ are Hölder continuous functions on [-1, 1].

It follows from (3.7), (3.8) that

$$q_k(\zeta) = \frac{Q_k(\beta)}{\sqrt{1-\beta^2 s'(\beta)}},$$
$$\{Q_1(\beta), Q_2(\beta)\} = \sum_{k=1}^{2}\{-1, \mu_k\}\Omega_k(\beta) \tag{4.2}$$
$$\Omega_k(\beta) = (-1)^k\frac{Q_2(\beta)+\varepsilon_k Q_1(\beta)}{\mu_2-\mu_1}, \varepsilon_1 = \mu_2, \varepsilon_2 = \mu_1$$

The asymptotes of the Cauchy integral at the ends are determined by the equalities [9]

$$f(z) = \frac{1}{2\pi i}\int_{\Gamma}\frac{\omega(\zeta)d\zeta}{(\zeta-c)^{\gamma}(\zeta-z)} = \begin{cases} \frac{e^{i\pi\gamma}\omega(a)}{2isin\pi\gamma}(z-a)^{-\gamma}+f_1(z), c=a \\ -\frac{e^{i\pi\gamma}\omega(b)}{2isin\pi\gamma}(z-b)^{-\gamma}+f_2(z), c=b \end{cases}$$

$$\lim_{z\to a} f_1(z)(z-a)^{\gamma}=0,\ \lim_{z\to b} f_2(z)(z-b)^{\gamma}=0, 0\le \Re\gamma<1 \tag{4.3}$$

where a is the beginning and b is the end of the arc L.

Taking into account (3.15) we represent the main asymptotes of the generalized Cauchy integral (3.2) in the form:

$$\Phi_k(z_k) = \frac{i\Omega_k(\pm1)}{2\sqrt{2\zeta_k'(\pm1)}}(\pm c_k \mp z_k)^{-\frac{1}{2}}, \zeta_k'(\beta) = \frac{d\zeta_k}{d\beta} \tag{4.4}$$

where the upper sign corresponds to the end of the crack $c_k = b_k$, the lower one to the beginning $c_k = a_k$, $b_k = \Re\, b + \mu_k \Im\, b$, $a_k = \Re\, a + \mu_k \Im\, a$.

Let

$$z-a = e^{re^{i\theta_a}}; z-b = e^{re^{i\theta_b}};\ 0\le\theta_a,\theta_b\ge\pi, \tag{4.5}$$

then (Fig. 2)

$$\begin{aligned} z_k - a_k &= \Re(z-a) + \mu_k\Im(z-a) = r(cos\theta_a + \mu_k sin\theta_a) \\ z_k - b_k &= r(cos\theta_b + \mu_k sin\theta_b) \end{aligned} \tag{4.6}$$

Taking into account (3.18) and (3.17) we obtain the main asymptotes for the stresses at the tips. The Eq. (2.2) yields

$$\{\sigma_{11},\sigma_{12},\sigma_{22}\} = \frac{1}{\sqrt{r}}\Re\sum_{k=1}^{2}\{\mu_k^2, -\mu_k,\ 1\}\frac{i\Omega_k(\pm1)}{2\zeta_k'(\pm1)}(\mp\, cos\theta_c \mp \mu_k sin\theta_c)^{-\frac{1}{2}} \tag{4.7}$$

where the upper sign corresponds to the tip $c = a$, the lower one to the tip $c = b$.

In order to compute the SIF K_I and K_{II} at the crack tips we find

$$\sigma_n = \sigma_{11}cos^2\psi + \sigma_{12}sin2\psi + \sigma_{22}sin^2\psi,\ \tau_{ns} = \sigma_{12}cos2\psi + \frac{\sigma_{22}-\sigma_{11}}{2}sin2\psi \tag{4.8}$$

where ψ is the angle between the normal vector and the axis Ox_1.

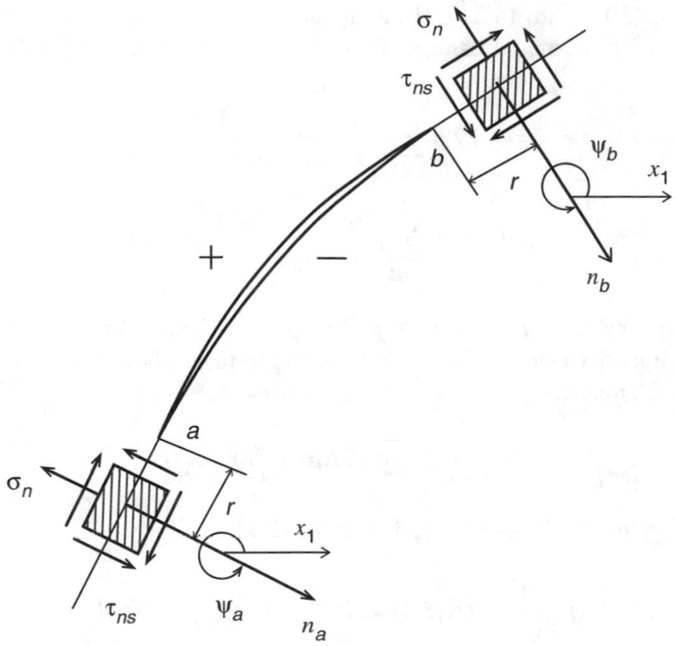

Fig. 2 Crack and stress near the tips

Taking into account the expressions for the stresses (2.2) we have

$$
\begin{aligned}
\sigma_n &= 2\Re \sum_{k=1}^{2} a_k^2(\psi)\Phi_k(z_k) \\
\tau_{ns} &= 2\Re \sum_{k=1}^{2} a_k b_k(\psi)\Phi_k(z_k) \\
b_k(\psi) &= -\mu_k \sin\psi - \cos\psi = \frac{da_k(\psi)}{d\psi}
\end{aligned}
\tag{4.9}
$$

In order to evaluate the main asymptotes of the stresses σ_n and τ_{ns} on the crack we use the formulas (4.2), where the angles ψ at every tip coincide with the angles of the corresponding normal vector ψ_a and ψ_b (Fig. 2).

In this case, we have

$$
z_k - a_k = -ra_k(\psi_a), z_k - b_k = -ra_k(\psi_b) \tag{4.10}
$$

Substituting (3.17) into (4.2) and taking into account (4.3) we obtain the main asymptotes of the normal stress σ_n and in-plane shear τ_{ns} at the tips of the crack

$$\begin{aligned} \sigma_n &= \mp 2\Re \sum_{k=1}^{2} \frac{\Omega_k(\pm 1) a_k(\psi_c)}{2\sqrt{2rs'(\pm q)}} + O(1) \\ \tau_{ns} &= \mp 2\Re \sum_{k=1}^{2} \frac{\Omega_k(\pm 1) b_k(\psi_c)}{2\sqrt{2rs'(\pm q)}} + O(1) \end{aligned} \tag{4.11}$$

Here, the upper sign also corresponds to the tip $c = b$ the lower one to the tip $c = a$. The stress intensity factors of normal break and in-plane shear at the crack tips are defined by the relations [10, 11].

$$K_I = \lim_{n\to 0} \sqrt{2\pi r}\sigma_n, K_{II} = \lim_{n\to 0} \sqrt{2\pi r}\tau_{ns} \tag{4.12}$$

Hence, using the relations (3.16) and (4.4) we obtain

$$\begin{aligned} K_I &= \mp \sqrt{\frac{\pi}{s'(\pm 1)}} \{Q_1(\pm 1) cos(\psi_c) + Q_2(\pm 1) sin(\psi_c)\} \\ K_{II} &= \mp \sqrt{\frac{\pi}{s'(\pm 1)}} \{-Q_1(\pm 1) sin(\psi_c) + Q_2(\pm 1) cos(\psi_c)\} \end{aligned} \tag{4.13}$$

and

$$\begin{aligned} Q_1(\pm 1) &= \pm \sqrt{\frac{s'(\pm 1)}{\pi}} \{K_I \sin\psi_c + K_{II} \cos\psi_c\} \\ Q_2(\pm 1) &= \mp \sqrt{\frac{s'(\pm 1)}{\pi}} \{K_I \cos\psi_c - K_{II} \sin\psi_c\} \end{aligned} \tag{4.14}$$

The Eq. (4.5) yields the expression

$$K_I - iK_{II} = \mp \sqrt{\frac{\pi}{s'(\pm 1)}} \{Q_2(\pm 1) + iQ_1(\pm 1)\} e^{i\psi_c} \tag{4.15}$$

where the upper sign corresponds to the tip $c = b$ the lower one to the tip $c = a$.

Let us consider two states of the deformed solid, weakened by the crack. Let us suppose that $\sigma_{ij}^{(0)}, e_{ij}^{(0)}, u_i^{(0)}$ are the components of the tensors of stresses, the deformation and displacement vector in an initial state of the solid "0", $\sigma_{ij}^{(1)}, e_{ij}^{(1)}, u_i^{(1)}$ are the corresponding values at the state "1". We obtain the increment $\Delta\Sigma$ for the part of the two-side surface of the crack. In the assumption of the absence of the bulk forces we write the expression for the increment of the inner energy in the case of the transition of the anisotropic solid from the state "0" into the state "1".

This increment $\Delta A_{\Delta\Sigma}$ determines the energy flow in the breaking $\Delta\Sigma$. Hence, we obtain [11]

$$\Delta A_{\Delta\Sigma} = \frac{1}{2}\int_{\Delta\Sigma_1} \sigma_{ij}^{(0)} n_j [u_i^{(1)}] ds, \; [u_i] = u_i^+ - u_i^- \tag{4.16}$$

Here, the sign $\pm$ corresponds to the limit values of $u_i^{(1)}$ on the double-side surface $\Delta\Sigma$. Th integral is taken at one side of the surface

$$\gamma(\Delta\Sigma_1 + \Lambda\Sigma_2) = -\Delta A_{\Lambda\Sigma} \tag{4.17}$$

where γ is the density of the energy surface.
The crack is expanded on the tangential on a small distance $h = \Delta\, l$ and its tip c takes a location c'. Then, (4.7) yields

$$\Delta A_{\Delta l} = \frac{1}{2}\int_0^h (X_{1n}\Delta u_1 + X_{2n}\Delta u_2) dr \tag{4.18}$$

where $\Delta\, u_1$, $\Delta\, u_2$ are the displacements of the notches of the extended crack at cc'. We obtain at the neighborhood of c'

$$\int \Phi_k(z_k) = \varphi_k(z_k) = \mp \frac{i\Omega(\pm 1)}{\sqrt{2\zeta_k'(\pm 1)}} \sqrt{\mp\left(z_k - c_k'\right)} \tag{4.19}$$

where, as always, the upper sign is related to the end of the expanded crack $c' = b'$, lower to the begin of $c' = a'$ (Fig. 3)

$$\begin{aligned} \Delta u_1 &= 2\sqrt{\frac{2(h-r)}{s'(1)}} s_{11}\{\, Q_2\Im(\mu_1 + \mu_2) - Q_1\Im(\mu_1\mu_2)\} \\ \Delta u_2 &= 2\sqrt{\frac{2(h-r)}{s'(1)}} s_{22}\{\, Q_1\Im\left(\frac{\mu_1 + \mu_2}{\mu_1\mu_2}\right) - Q_2\Im\left(\frac{1}{\mu_1\mu_2}\right)\} \end{aligned} \tag{4.20}$$

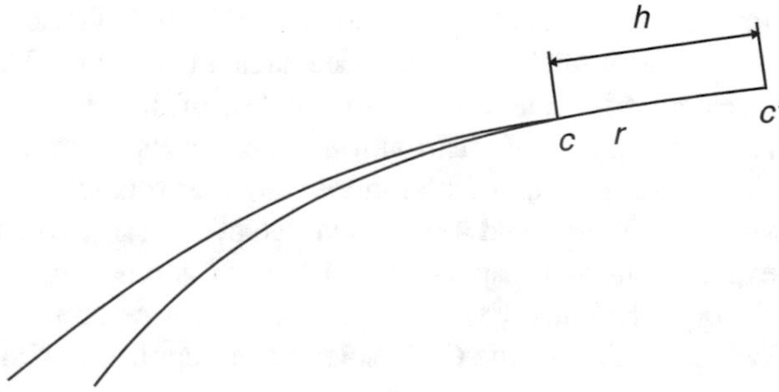

Fig. 3 States "0" and "1" into evaluation energy at the tips

The stress vector projection has the form

$$X_{1n} = \frac{1}{\sqrt{2rs'(1)}} \Re \sum_{k=1}^{2} \mu_k \Omega_k(1) = \frac{Q_2(1)}{2rs'(1)},$$
$$X_{2n} = -\frac{1}{\sqrt{2rs'(1)}} \Re \sum_{k=1}^{2} \Omega_k(1) = -\frac{Q_2(1)}{2rs'(1)} \tag{4.21}$$

Using the formulas (4.11), (4.12), derive from (4.9)

$$\Delta A_{\Delta l} = \frac{\pi h}{2s'} s_{11}(\alpha_{11} Q_1^2 - 2\alpha_{12} Q_1 Q_2 + \alpha_{22} Q_2^2) \tag{4.22}$$

Finally, taking into account (4.6) we obtain

$$\Delta A_{\Delta l} = \frac{h}{2} s_{11}(\beta_{11} K_I^2 - 2\beta_{12} K_I K_{II} + \beta_{22} K_{II}^2) \tag{4.23}$$

where

$$\begin{aligned}
\alpha_{11} &= \Im[(\mu_1 + \mu_2)(\overline{\mu_1 \mu_2})] \\
\alpha_{12} &= \Im[(\mu_1 \mu_2)] \\
\alpha_{11} &= \Im[(\mu_1 + \mu_2)] \\
\beta_{11} &= \alpha_{11} sin^2\psi - \alpha_{12} sin2\psi + \alpha_{22} cos^2\psi \\
\beta_{12} &= (\alpha_{11} - \alpha_{22}) sin2\psi - 2\alpha_{12} cos2\psi \\
\beta_{22} &= \alpha_{11} cos^2\psi + \alpha_{12} sin2\psi + \alpha_{22} sin^2\psi
\end{aligned} \tag{4.24}$$

5 Results and Conclusion

We consider numerical experiments in a finite plane with 50–100 random normally distributed cracks, concentrated in a bounded area in a plane. The cracks are obtained using Delaney triangulation with the centers of the triangulation ordered according to the normal law of distribution. The cracks were formed by the polynomial of the third degree and with random angle of rotation. The experiment has been repeated 5–10 thousand times. On its every stage we obtained the mathematical expectations and dispersions for the stress intensity factors and the energy flows at every crack tips. Experiments were performed for various variants of the mechanical loads. The Monte Carlo scheme was applied in simulations. The results are represented in the form of the histogram of the normalized frequencies for the obtained sample x

$$\nu_n = \frac{\nu_i}{S}, \quad S = \sum_i \nu_i \Delta x \tag{5.1}$$

and for the probability density function

$$f(x, \mu, \sigma) = \frac{1}{\sigma\sqrt{2\pi}} exp\left[-\frac{(x-\mu)^2}{2\sigma^2}\right] \tag{5.2}$$

μ—mean, σ—standard deviation of sample x.

Using the described above methods we take the stresses at the infinity as $\sigma_{22} = 1\frac{N}{m^2}$; $\sigma_{11} = \sigma_{12} = p_n = 0$;

The deformation coefficients of the anisotropic plate [1] is $s_{11} = s_{22} = 14,362 \times s_0$; $s_{12} = -4,2625 \times s_0$; $s_{16} = -s_{26} = -1,7754$; $s_{66} = 41.35 \times s_0$; $s_0 = 10^{-12}\frac{N}{m^2}$

As a final conclusion we can say that the normal law of the distribution of the cracks in the plane yields the normal law of the distribution of the parameters of distraction (Figs. 4, 5, 6 and 7).

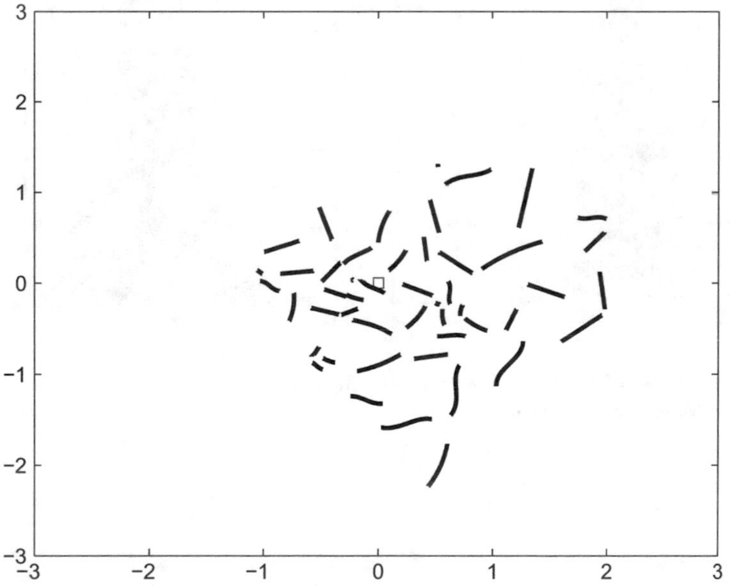

Fig. 4 An example of the configuration of 50 cracks

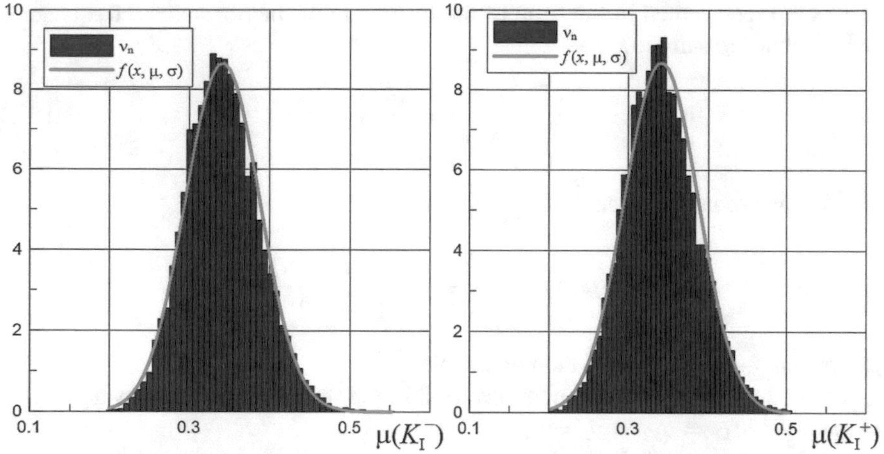

Fig. 5 Distribution of the mean values of the stress intensity factor K_I. The sign $-$ corresponds to the tip a of the crack, sign $+$ to the tip b

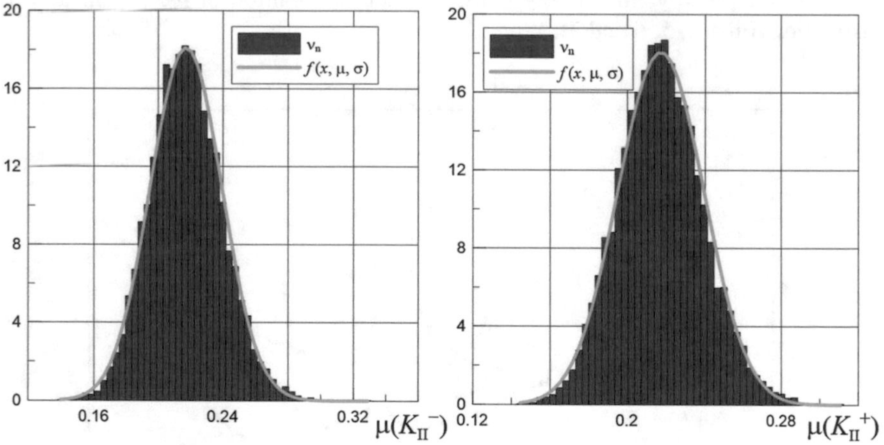

Fig. 6 Distribution of mean values of stress intensity factor K_{II}. The sign $-$ corresponds to the tip a of the crack, sign $+$ to the tip b

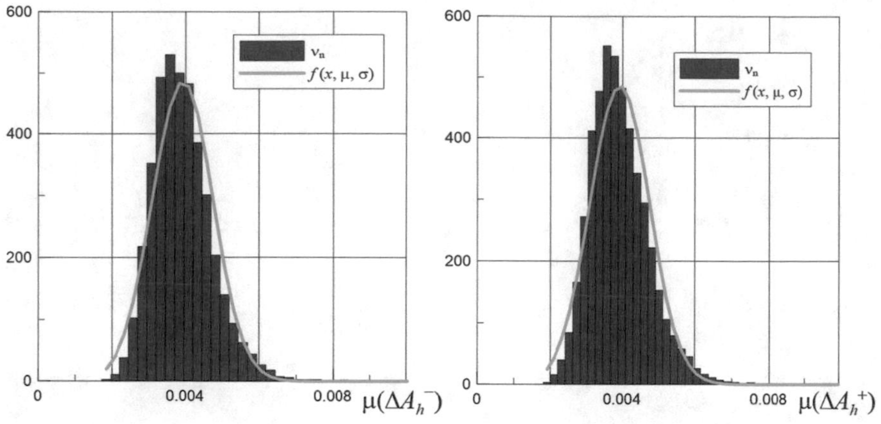

Fig. 7 Distribution of mean values of energy flows ΔA_h. The sign $-$ corresponds to the tip a of the crack, sign $+$ to the tip b

References

1. E.I. Grigolyuk, L.A. Filshtinsky, *Regular Piece-Homogeneous Structures with Defects* (Fiziko-Matematicheskaja Literatura, Moscow, 1994); (In Russian)
2. M. Denda, Mixed mode I, II and III analysis of multiple cracks in plane anisotropic solids by the BEM: a dislokation and point force approach. Eng. Anal. Bound. Elem. **25**, 267–278 (2001)
3. T. Ikeda, M. Nagai, K. Yamanaga, N. Miyazaki, Stress intensity factor analyses of interface cracks between dissimilar anisotropic materials using the finite element method. Eng. Fract. Mech. **73**, 2067–2079 (2006)
4. E. Rejwer, L. Rybarska-Rusinek, A. Linkov, The complex variable fast multipole boundary element method for the analysis of strongly inhomogeneous media. Eng. Anal. Bound. Elem. **43**, 105–116 (2014)
5. J. Wang, S.L. Crouch, S.G. Mogilevskaya, A fast and accurate algorithm for a Galerkin boundary integral method. Comput. Mech. **37**, 96–109 (2005)
6. M.D. Snyder, T.A. Cruse, Boundary-integral equation analysis of cracked anisotropic plates. Int. J. Fract. **11**, 315–328 (1975)
7. S.G. Lekhnitskii, *Anisotropic Plates*, vol. 355 (Gordon and Breach Science Publishers, New York, 1968)
8. L.A. Filshtinskii, Elastic equilibrium of anisotropic medium, weakened bu random curveliniar cracks. Limiting process to isotropic medium. Izv. ANSSSR: Mech. Tw. Tela **5**, 91–97 (1976)
9. N.I. Muskhelishvili, *Singular Integral Equations*, 3rd edn. (Nauka, Moscow, 1968); (in Russian) English transl. of the 1st ed.: P. Noordhoff N.V., Groningen, 1946
10. G.P. Cherepanov, *Mechanics of Brittle Distruction* (Nauka, Moscow, 1974), p. 640
11. V.V. Panasyuk, M.P. Savruk, A.P. Dacyshan, *Stress Distribution at the Neighbourhood of the Cracksin Plates and Shells*, vol. 444 (Naukova Dumka, Kyiv, 1976)

Perturbative Expansions and Critical Phenomena in Random Structured Media

Simon Gluzman and Dmitry A. Karpeyev

Abstract We present constructive solutions for the effective properties for three problems in the field of random structured media. They are all based on truncated series and on a constructive investigation of their behavior near divergence points where the physical percolation or phase transitions occur. (1) Effective conductivity of 2D conductors with arbitrary contrast parameters is reconstructed from the expansion at small concentrations and of the critical behavior at high concentrations. (2) Effective shear modulus of perfectly rigid spherical inclusions randomly embedded into an incompressible matrix is reconstructed given its expansion at small concentrations and critical behavior. In addition, the critical index S of super-elasticity is estimated. (3) We also employ a truncated Fourier expansion to study spontaneous directional ordering in models of planar fully-connected suspensions of active polar particles. The main result is the discovery of a discontinuous, abrupt transition from an ordered to a disordered state. It is a macroscopic effect caused by a mesoscopic self-quenching noise. The relaxation time remains finite at the critical point, therefore the effect of self-quenching is to strongly suppress the critical slowing down and improve the reaction time to external stimuli.

Keywords Effective conductivity • Effective shear modulus • Active suspensions • Discontinuous transition

---

S. Gluzman (✉)
Department of Mathematics, Pennsylvania State University, University Park, State College, PA 16802, USA
Present address: 3000 Bathurst St, Apt.606, Toronto ON, Canada M6B 3B4
e-mail: simon.gluzman@gmail.com

D.A. Karpeyev
Computation Institute, University of Chicago, 5735 S. Ellis Avenue, Chicago, IL 60637, USA

Mathematics and Computer Science Division, Argonne National Laboratory, 9700 S. Cass Avenue, Argonne, IL 60439, USA
Present address: 2656 Bryant Avenue, Evanston, IL 60201, USA
e-mail: karpeev@mcs.anl.gov

P. Drygaś, S. Rogosin (eds.), *Modern Problems in Applied Analysis*,
Trends in Mathematics, https://doi.org/10.1007/978-3-319-72640-3_9

Mathematics Subject Classification (2010) Primary 40C, 41-06; Secondary 74B, 74Q, 92B

The three problems considered in the paper belong to the field of random structured media undergoing a phase transition with changing strength of the disorder or noise. Such problems are treated by application of some perturbation theory in terms of the corresponding small parameters, when only a few starting terms in expansions of the effective properties are available. In order to extend the region of applicability of the corresponding series one has to apply some re-summation technique. The self-similar renormalization appears to be uniquely suited for all three problems with necessary modifications taking into account the specifics.

Property of self-similarity is the central concept of the approach and represents the group property of a function conserving its form under the change of its variable. It is pertinent to the autonomous dynamical systems, expressing the group property of motion. Similar relations appear in the renormalization group approach of quantum field theory. The concept of self-similarity re-appears in the context of fractals. In particular cases, this variable can be time, as for dynamical systems, momentum, as for field theory, or space scale, as for fractals. Most generally such a variable can be the approximation number, playing the role of effective time. Then motion occurs in the space of approximations, and the property of self-similarity is applied constructively to extend the low-order formulae to make them applicable to arbitrary values of parameters.

1 2D Conductivity: Dependence on Contrast Parameter

Consider a classical problem of the effective conductivity of a 2D random composite. An accurate approximate formula can be deduced for a 2D, two-component composite made from a collection of non-overlapping, identical, ideally conducting circular discs, embedded randomly in an otherwise uniform locally isotropic host. Consider the general situation, when contrast parameter

$$\rho = \frac{\sigma_i - \sigma_m}{\sigma_i + \sigma_m}, \tag{1}$$

enters the power series for conductivity explicitly. Usually for the matrix it is taken $\sigma_m = 1$, and for the inclusions $\sigma_i \equiv \sigma$, so that $|\rho| \leq 1$.

There are two different limit cases for the effective conductivity. In the limiting case of a perfectly conducting inclusions, the effective conductivity is expected to tend to infinity as a power-law, as the concentration of inclusions x tends to the maximal value $x_c = \frac{\pi}{\sqrt{12}}$ achievable for the random array,

$$\sigma(x) \simeq A(x_c - x)^{-s}. \tag{2}$$

The critical superconductivity index (exponent) s is believed to be close to $\frac{4}{3} \approx 1.3$ [23]. The critical amplitude A is an unknown non-universal parameter. From the phase interchange theorem [23] it follows that in two-dimensions, the superconductivity index s is equal to the conductivity index t.

On the other hand the following expansion in concentration x of the inclusions and contrast parameter ρ was obtained in[9],

$$\sigma(x, \rho) = 1 + a_1(\rho)x + a_2(\rho)x^2 + a_3(\rho)x^3 + \dots \tag{3}$$

The low-order coefficients depend only on ρ,

$$a_1(\rho) = 2\rho, \quad a_2(\rho) = 2\rho^2, \quad a_3(\rho) = 4.9843\rho^3 - 0.0688\rho^4 - 0.1463\rho^5 - 0.7996\rho^6. \tag{4}$$

Since there are two parameters x and ρ, we derive the final formula in two steps. First, let us guarantee the correct dependence on contrast parameter.

Let us stress that in 2D one has to respect celebrated Keller's phase-interchange relation [14, 23], valid for the general case of average conductivity of a statistically homogeneous isotropic random distribution of cylinders of one medium in another medium [14]. Since the dependence on the conductivity of inclusions and matrix is hidden within the contrast parameter, the phase interchange can be expressed even simpler as follows,

$$\frac{1}{\sigma(x, \rho)} = \sigma(x, -\rho), \tag{5}$$

One should try to develop an expression which satisfies (5) by design. The proper steps have to be taken to guarantee corresponding critical properties. One has then simply modify parameters of the solution to move away to non-critical situations. The simplest solution along these lines can be constructed in terms of the factor approximant [26]. They are particularly suited to include both types of critical behavior as $\rho \to -1$, and $\rho \to 1$, respectively,

$$\Sigma^*(x, \rho) = (B_1(\rho)x + 1)^{4/3}(B_2(\rho)x + 1)^{-4/3}. \tag{6}$$

Here $B_1(\rho) = \frac{3\rho}{4}$, $B_2(\rho) = -\frac{3\rho}{4}$, are defined from the asymptotic equivalence to the series (3). The symmetry dictated by (5) is self-evident, but the values of thresholds as $\rho \to -1$, and $\rho \to 1$, are wrong. Let us impose the conditions on correct threshold in both limits by means of the simple exponential form for the effective thresholds. Thus we arrive to the following re-summed expression

$$\Sigma^{**}(x, \rho) = (B_1^*(\rho)x + 1)^{4/3}(B_2^*(\rho)x + 1)^{-4/3}, \tag{7}$$

where $B_1^*(\rho) = \frac{3\rho}{4e^{0.385406\rho}}$, $B_2^*(\rho) = -\frac{3\rho}{4e^{-0.385406\rho}}$. Let us find a multiplicative correction to (7), expressed as the Pade approximant We thus develop the following corrected factor approximant,

$$\Sigma_{1c}^{**}(x,\rho) = \Sigma^{**}(x,\rho)\frac{1+vx}{1-vx}, \tag{8}$$

where $v = v(\rho)$ and

$$v(\rho) = 0.5\left(-e^{-0.385406\rho}\rho - e^{0.385406\rho}\rho + 2\rho\right). \tag{9}$$

The latter expression is obtained from the asymptotic equivalence with series (3). The form of correction is additionally motivated by the celebrated Clausius-Mossotti (CM) formula, $\sigma_{CM}(x,\rho) = \frac{1+\rho x}{1-\rho x}$ [4]. CM is valid for small concentrations but respects the phase interchange symmetry. Formula (8) respects this symmetry as well.

Assuming even more general form for the correcting Pade approximant, we obtain another more accurate formula.

$$\Sigma_{2c}^{**}(x,\rho) = \Sigma^{**}(x,\rho)\frac{w_1(\rho)\rho x + w_2(\rho) r^3 x^2 + 1}{-w_1(\rho)\rho x - w_2(\rho)\rho^3 x^2 + 1}, \tag{10}$$

where

$$w_1(\rho) = 1 - \cosh(0.385406\rho), \quad w_2(\rho) = -\frac{0.375\sinh(0.770811\rho)}{\rho}. \tag{11}$$

The phase-interchange symmetry is again preserved.

Consider now the simplest possible dependence of the effective thresholds leading to the correct value of x_c as $\rho \to -1$ and $\rho \to 1$ respectively. Such a dependence follows from the CM formula, $B_1^*(\rho) = \frac{\rho}{x_c}$, $B_2^*(\rho) = -\frac{\rho}{x_c}$. Using the same approach as above we obtain much simpler expression for the effective conductivity,

$$\Sigma_s^{**}(x,\rho) = \frac{\left(\frac{\rho\sqrt{3}rx}{\pi}+1\right)^{4/3}\left(\frac{\left(\pi-\frac{\rho}{\sqrt{3}}\right)rx}{\pi}+1\right)}{\left(1-\frac{\rho\sqrt{3}rx}{\pi}\right)^{4/3}\left(1-\frac{\left(\pi-\frac{\rho}{\sqrt{3}}\right)rx}{\pi}\right)}, \tag{12}$$

which gives results very close to the more sophisticated (10). Moving singularity to the non-physical values of x allows to preserve the form typical for critical regime for all values of ρ, but also suppresses the singularity. Assumptions leading to formula (12) are most intuitive and can serve as a guide in generally more complicated situations. It is also significant that the formulae suggested above do agree with upper and lower Hashin-Strickman bounds given by Eq. (7.10) from [16].

2 3D Elasticity, or High-Frequency Viscosity: Critical Index

Consider a problem of perfectly rigid spherical inclusions randomly embedded into an incompressible matrix. It is analogous to the problem of high-frequency effective viscosity of a suspension [3, 8]. Such composite has a Newtonian elastic behavior and the ratio of the effective shear modulus to that of the matrix is [8, 24],

$$\mu(x) = 1 + a_1 x + a_2 x^2 + O(x^3), \quad a_1 = 5/2, \quad a_2 = 5.0022. \tag{13}$$

It is expected that $\mu(x) \simeq A(x_c - x)^{-S}$, in the vicinity of the 3D-threshold. The "super-elasticity index" S is defined for super-rigid inclusions and is analogous to the superconductivity index s [6]. In the 2D case Bergman argued that $S = s$. Let us employ some special resummation technique uniquely suited for very short series and estimate the critical index S [25]. The technique revealed very good results for the conductivity critical index in 2D and 3D based on the short expansions in concentration. Let us apply the resummation technique to the inverse series (13),

$$\mu(x)^{-1} = M(x) \simeq 1 + \tilde{a_1} x + \tilde{a_2} x^2, \quad \tilde{a_1} = -5/2, \quad \tilde{a_2} = 1.2477. \tag{14}$$

Then one should construct two different approximants. For instance the simplest pair is given by the following low-order approximants with c being a control parameter,

$$M_1^*(x) = \left(\frac{c}{c - \tilde{a_1} x}\right)^c, \quad M_2^*(x) = 1 + \tilde{a_1} x \left[1 - \frac{\tilde{a_2} x}{\tilde{a_1}(1+c)}\right]^{-(1+c)}. \tag{15}$$

From the first-order approximant we estimate threshold as a function of c, $x_c(c) = \frac{c}{\tilde{a_1}}$. Naturally one would like to have the two approximants to differ from each other minimally. The minimal difference condition gives the condition on stabilizer c. It will be determined as a *minimizer* of the following expression

$$\left| 1 + \tilde{a_1} x_c(c) \left[1 - \frac{\tilde{a_2} x_c(c)}{\tilde{a_1}(1+c)}\right]^{-(1+c)} \right|. \tag{16}$$

We find numerically that $c = -1.57637$, leading to the critical index $S = -c$. Even more convincing is the corresponding value of threshold $x_c(c) = 0.630549$. While the correct threshold is expected to be 0.637. It gives more credence to the estimate of index S. What happens to the index estimate if we do know the correct value of threshold $x_c = 0.637$ in advance, and try to use the knowledge? Let us apply the transformation, $z = \frac{x}{x_c - x}$, to the original series (13), with the resulting series

$$\mu(z) = 1 + b_1 z + b_2 z^2 + O(z^3), \quad b_1 = 1.5925, \quad b_2 = 0.437278. \tag{17}$$

The set of approximations to $\mu(z)$ including the two starting terms from (17), leads to the simple expression for the renormalized quantity μ_1^*,

$$\mu_1^*(z) = \left(1 - \frac{b_1 z}{s_1}\right)^{-s_1} \Longrightarrow \left(\frac{s_1}{-b_1}\right)^{s_1} z^{-s_1} \; (z \to \infty). \tag{18}$$

The stabilizer s_1 should be negative, if we want to reproduce in the limit of $z \to \infty$, the correct power-low behavior of the effective shear modulus. For comparison one also needs to construct a different set of approximations. It can be accomplished simply by leaving the constant term from (17) outside the renormalization procedure. Then the second-order approximation appears as follows [25],

$$\begin{aligned} \mu_2^*(z) &= 1 + b_1 z[1 - \tfrac{b_2 z}{b_1(1+s_2)}]^{-(1+s_2)} \Longrightarrow \\ &(-\tfrac{b_2}{1+s_2})^{-(1+s_2)} \, b_1^{2+s_2} z^{-s_2} \; (z \to \infty). \end{aligned} \tag{19}$$

Demand now that both (18) and (19) have the same power-law behavior at $z \to \infty$, while $s_2 = s_1 \equiv c$. Requiring the convergence for the two available approximations in the form of the minimal-difference condition for critical amplitude, we obtain the condition on the *negative* stabilizer c. It is determined numerically from the *minimum* of the expression

$$\left\| \left[\left(\frac{-b_2}{1+c}\right)^{-(1+c)} b_1^{(2+c)} - \left(\frac{c}{-b_1}\right)^{c} \right] \right\|. \tag{20}$$

The nontrivial zero of (20) is located at the point $c = -1.60491$. The final formula which respects (13), has the following form:

$$\mu_2^*(z) = 1 + b_1 z \left[1 - \frac{b_2 z}{b_1(1+c)}\right]^{-(1+c)}. \tag{21}$$

Expressed in the original variable it gives the effective shear modulus

$$\mu^*(x) = 1 + \frac{1.5925 \left(\frac{0.637 - 0.546068x}{0.637 - x}\right)^{0.604905} x}{0.637 - x}, \tag{22}$$

for arbitrary concentration of particles and the critical index $S = -c = 1.60495$. Note also that a naive estimate with the simplest root approximant $\mu^*(x) = (1 + Bx)^{-S}$ gives a bit higher value $S = 1.66462$, with $B = -1.50184$.

All our estimates thus agree with the value of 1.7 quoted in [22]. Note that S predicted for the high-frequency viscosity of suspensions is smaller than 2, the value expected in the general case of 3D viscous suspensions [7]. The critical amplitude can be estimated $A = 0.47891$. The formula (22) can be expanded at small x,

$$\mu^*(x) = 1 + 2.5x + 5.0023x^2 + 9.39296x^3 + 16.9754x^4 + O(x^5).$$

The third-order coefficient a_3 equals 9.39296, in agreement with the most sophisticated estimate 9.09 ± 0.02 from [8]. The fourth-order coefficient is estimated as well.

Formula (22) can be compared with analytical expression from [8],

$$\mu_A(x) = 1 + \frac{5\left(0.63x^2 + 1.0009x + 1\right)x}{2\left(-0.63x^3 - 1.0009x^2 - x + 1\right)}.$$

There is a good agreement up to $x = 0.45$. The latter formula definitely fails in the region of high-concentrations, since it predicts an incorrect threshold and index, while (22) captures this region close to x_c and predicts index S.

3 Spontaneous Ordering in Suspensions of Active Particles

Active Brownian particles have the ability to generate a flow field, which in turn can influence their motion. In addition, many active particles are polar, such as, for example biological agents or chemically-driven colloids. They have a distinct body asymmetry, an axis defining their preferred direction of motion (the head-tail axis). The polarity of particles introduces a distinct orientation, which in two dimensions is entirely determined by a single angular variable [19]. An important feature of most active matter systems is the presence of non-negligible random fluctuations in the motion of individual active units. Quantification of this apparent randomness is considered below at the level of Fokker-Planck equations for the evolution of the probability densities.

Microscopic equations of motion of active interacting particles, such as swimming bacteria or schooling fish, are effectively intractable, having to describe the dynamics with up to $\mathcal{O}(10^6)$ degrees of freedom. Instead, we propose a kinetic theory, the Generalized Fokker-Planck Equation (GFPE), which seeks to systematically incorporate collisions and ordering effects specific to polar particles beyond the classical mean-field approximation to the microscopic equations.

The GFPE, like many other kinetic theories, is not derived rigorously but rather by plausible arguments, leading to an interplay of the (angular) diffusion and inelastic collisions—the main physical effects of interest in this model. Competition of alignment effected by collisions [2, 20], and disalignment due to diffusion is expected to lead to a spontaneous directional ordering at the critical point where collisions overcome diffusion. Here we consider several models of this type in 2D with particle position $\boldsymbol{r} = (x, y)$ and orientation $\boldsymbol{\tau}$ that can be characterized by a single angle θ as in $\boldsymbol{\tau} = e^{i\theta} = (\cos\theta, \sin\theta)$. In this setting the probability density is a function of the in-plane angle and position $P(\boldsymbol{r}, \theta, t)$.

GFPE is derived from the minimal Model 0 (see Eq. (23)) initially proposed by [1] in a different context and analyzed in [5]. This model describes a *fully-connected* suspension of polar (or oriented) particles in the plane. That is, our model describes a suspension where all particles interact with all other particles and governs only the angular part of the single-particle distribution function $P(\theta, t)$, since the spatial position is irrelevant in the situation of full connectivity. The relevant physical models are, for example, suspensions of swimming bacteria and mixtures of microtubules interacting by means of molecular motors. While the assumption of full connectivity implies spatial homogeneity and limits its applications to real dynamics, these models are attractive since they remain analytically and/or numerically tractable, while spatially-heterogeneous models are rarely so. At the same time, fully-connected models (see e.g., [18]) can still capture a nontrivial aspect of the real physics. In the fully-connected situation the movement of particles reduces to changes in orientations, which has the diffusive and steric parts. In this context, the angular diffusion, characterized by coefficient D, can be viewed as modeling local noise as well as the coarse-grained fluctuations not captured by the mean long-range spatial interactions. For example, in the case of swimming bacteria the noise is a model of intrinsic stochastic effects, such as bacterial tumbling, as well as other non-systemic effects that are not captured by our extended mean-field model. These and other weaker effects that cumulatively act as additive Gaussian white noise. Steric interactions are modeled by a binary integral operator, which describes the inelastic collisions of collocated particles that bring each colliding particle pair into complete alignment.

A central question in the study of models of collective motions is the presence and the nature of phase transitions, which are typically characterized by the order parameter $\overline{\boldsymbol{\tau}}$—the suitably-defined local mean orientation. In the disordered phase $|\overline{\boldsymbol{\tau}}| \sim 0$, while in the strongly ordered regime $|\overline{\boldsymbol{\tau}}| \sim 1$.

GFPE Models Model 0 is perhaps the simplest translationally-invariant model of interacting polar particles in a 2-dimensional domain with inelastic collisions. This description is particularly suitable for suspensions of particles in a thin liquid layer and is borrowed from [1]. Here orientation $\boldsymbol{\tau}$ is uniquely defined by an angle θ, and the probability of such an orientation at time t is $P(\theta, t)$ governed by GFPE:

$$\frac{\partial P}{\partial t}(\theta, t) = \frac{\partial^2}{\partial \theta^2}\left(DP(\theta, t)\right) + \int_{-\pi}^{\pi} P\left(\theta - \frac{\theta'}{2}, t\right) P\left(\theta + \frac{\theta'}{2}, t\right) d\theta' - P(\theta, t). \tag{23}$$

Here D being the constant coefficient of angular diffusion normalized by the collision rate. Collisions themselves are encoded by the integral and linear on P term,

$$\begin{gathered}\int_{-\pi}^{\pi} P\left(\theta - \frac{\theta'}{2}, t\right) P\left(\theta + \frac{\theta'}{2}, t\right) d\theta' - P(\theta, t) = \\ \int_{-\pi}^{\pi} P\left(\theta - \frac{\theta'}{2}, t\right) P\left(\theta + \frac{\theta'}{2}, t\right) d\theta' - \int_{-\pi}^{\pi} P(\theta, t) P(\theta + \theta', t) d\theta'.\end{gathered} \tag{24}$$

The integral term is due to the increase in the probability of orientation θ as a result of a collision between two particles with pre-collision orientations $\theta + \theta'$ and $\theta + \theta'$. The linear term is due to the decrease in the same probability due to a collision between a particle oriented along θ with a differently-oriented particle $\theta + \theta'$. The usual normalization of total probability $\int P(\theta', t)d\theta' = 1$ is always respected, and the latter integral reduces to $P(\theta, t)$.

Model 0 assumes that rotational fluctuations on particle orientation at different orientations θ are decorrelated [2, 11]. There are experimental reasons to believe, however, that in a locally-ordered configuration local fluctuations become relatively weaker, requiring a stronger perturbation to break particle alignment [11]. Thus, it is more realistic to model orientational fluctuations as a self-quenching noise: a noise that decays with the increasing local ordering. We incorporate these ideas about fluctuations in our Model 1 where the self-quenching fluctuations are modeled as Gaussian white noise, but with its magnitude inversely related to a measure of ordering. It is convenient and natural to select mean orientation $\overline{\boldsymbol{\tau}}$ as the order parameter. Identifying orientation with a complex scalar $\boldsymbol{\tau} = e^{i\theta}$ we have

$$\overline{\boldsymbol{\tau}} = \int \boldsymbol{\tau} P(\theta)d\theta = \int e^{i\theta} P(\theta)d\theta.$$

This can be identified in terms of the Fourier series expansion of probability is $P(\theta) = \sum_k P_k e^{ik\theta}$, which has complex-conjugate coefficients $P_{-k} = P_k^*$ since $P(\theta)$ is real:

$$\overline{\boldsymbol{\tau}} = \int_{-\pi}^{\pi} e^{i\theta} P(\theta)d\theta = P_{-1} = P_1^*. \tag{25}$$

Our collisional model doesn't have a quantity that can be easily identified with the strength of alignment, except, perhaps the total collision rate implicitly present in the normalized maximum diffusion D. We can, however, consider a parameterized family of diffusion coefficients that depend on the order parameter, and also investigate the dependence of phase transitions on the parameterization.

To this end, we set general expression for the self-quenching diffusion coefficient $\overline{D}$ to be

$$\overline{D}(\overline{\boldsymbol{\tau}}) = D\left(1 - \kappa |P_1|^{\alpha}\right), \tag{26}$$

where D is the maximal diffusion strength normalized by the collision intensity, and $1 \le \alpha \le 2$ and $\kappa \ge 0$ are parameters.

Our main goal is to investigate the stationary behavior and phase transitions in the resulting Model 1

$$\frac{\partial P}{\partial t}(\theta, t) = \frac{\partial^2}{\partial \theta^2}\left(\overline{D}(\overline{\boldsymbol{\tau}})P(\theta, t)\right) + \int_{-\pi}^{\pi} P\left(\theta - \frac{\theta'}{2}, t\right) P\left(\theta + \frac{\theta'}{2}, t\right) d\theta' - P(\theta, t). \tag{27}$$

Specifically, we use two values of $\alpha = 1, 2$ and arbitrary κ. This range of α is representative of the full spectrum of behavior—continuous and discontinuous transitions. In particular, we analyze the endpoints of the range. For $\alpha = 1$ the transition becomes discontinuous in the critical point, in a marked difference with the case of $\alpha = 2$, with order parameter emerging continuously at the critical point. The latter remains qualitatively similar to Model 0 with a modified critical amplitude. These results allow us to conjecture that for some α^* between 1 and 2 there is a change from continuous to discontinuous behavior.

Phase transitions of GFPE are studied by examining its stationary solutions, which in general can only be obtained approximately. Since this approximation method is of independent interest, we develop it in some generality using Model 0. As essentially exact results for that model are available in [5], we can evaluate the approximation method in some detail. We seek a solution in terms of a truncated Fourier expansion of the form

$$P^{(k)}(\theta) = \frac{1}{2\pi} + a\cos(\theta) + b\cos(2\theta) + \cdots + d\cos(k\theta). \tag{28}$$

Observe that $P^{(0)} = P_0 = \frac{1}{2\pi}$ is the isotropic solution that is only expected to exist for sufficiently strong diffusion D. For D below a certain *critical diffusion strength* D_c the steady state becomes anisotropic. The basic symmetry $\theta \mapsto -\theta$ of the stationary GFPE implies that any anisotropic solution is symmetric about any of its local maxima (peak). In the ansatz above we assume a peak at $\theta = 0$, then the same symmetry explains the absence of $\sin(k\theta)$ terms in the above expansion.

3.1 Model 0: Stationary States with Additive Noise

Denote by E the stationary GFPE operator—the right-hand side of (23):

$$E[P](\theta) = D\frac{\partial^2}{\partial\theta^2}P(\theta,t) + \int_{-\pi}^{\pi} P\left(\theta - \frac{\theta'}{2}, t\right) P\left(\theta + \frac{\theta'}{2}, t\right) d\theta' - P(\theta,t). \tag{29}$$

Define, further, $R^{(k)}(\theta) = |E[P^{(k)}](\theta)|$ the residual at θ resulting from the substitution of approximant $P^{(k)}$. The expansion coefficients $a, b, c, \ldots$ are now determined using the following mixture of (a) Galerkin conditions, and (b) pointwise collocation conditions applied to R. Galerkin conditions require that $R^{(k)}$ be L^2-orthogonal to the corresponding low-order Fourier monomials $\cos(\theta)$ and $\cos(2\theta)$:

$$\int_{-\pi}^{\pi} R^{(k)}(\theta)\cos(m\theta)d\theta = 0. \tag{30}$$

Applying these conditions results in equations from which the corresponding Fourier coefficients $a, b, c, \ldots$ can be determined. Expansions (28) with the coefficients determined this way will be called *Fourier Polynomials* (FP) approximating the stationary solution. We also know that for $D < D_c = \frac{4}{\pi} - 1$ the stationary solution of Model 0 has a single maximum and is, therefore monotone on $[0, \pi]$ and $[-\pi, 0]$. We use this feature along with the positivity of P as an important criterion when evaluating the accuracy of our approximations.

Since the Fourier amplitudes in [5] are obtained from Galerkin conditions, we know that with the increasing order FP expansions converge in the L^2-norm to a stationary solutions of Model 0, implying that $||R^{(k)}||_{L^2} \rightarrow 0$. For finite k, however, the properties of the FP expansions can be far from the limit. While the Galerkin condition captures the behavior near the peak $\theta = 0$ well, and may faithfully reproduce the *mean* properties of P, FP approximations can produce spurious *pointwise* behavior away from $\theta = 0$. An example of this is a loss of monotonicity on $[0, \pi]$ resulting in multimodal distributions with multiple maxima. More significantly, however, FP approximations can have *negative* values, violating the probabilistic interpretation of P. To rectify this we impose a number of what is known in the numerical analysis literature *collocation* conditions—the vanishing of the residual at a number of suitably selected points (locations). Applying such collocation conditions at a number of select points θ_i we obtain:

$$R^{(k)}(\theta_i) = 0. \tag{31}$$

This approach is inspired by the approach to cross-over phenomena in [12, 26], where it was shown that imposing collocation conditions at or near the boundary of the domain results in an improved finite-order approximation, which we call the *Improved Fourier Polynomial* (IFP). Although in the present context the periodic domain does not have a boundary, as indicated above, pointwise behavior of FPs suffers away from the peak location $\theta = 0$. We therefore select $\theta_1 = \pi$ as the analog of the boundary and apply the first collocation condition there. As a rule, for our k-th order IFP approximation IFP_k we will determine all but the highest-order coefficient from the Galerkin conditions. In most cases, in order to restore the desirable monotonicity and positivity properties, it is sufficient to use a single collocation condition at $\theta_1 = \pi$ when determining the k-th order coefficient ($k > 2$). Thus, while FP_2 is identical to IFP_2, for high orders IFP_k differ from FP_k and are superior to them, as we now illustrate.

We start with the second order FP_2

$$R^{(2)}(\theta) = -\frac{a\,(\pi(4b + 3D + 3) - 12)}{3\pi}\cos(\theta) - \left(-\pi a^2 + 4bD + b\right)\cos(2\theta) + \ldots.$$

The first and second order Galerkin conditions are equivalent to setting the first and second order coefficients of $R^{(2)}$ to zero, recursively obtaining a and b as functions of the noise strength D:

$$a = \left(\frac{3(4D + 1)(D_c - D)}{4\pi}\right)^{1/2}, \; b = \frac{\pi a^2}{4D + 1}. \tag{32}$$

Thus FP_2 calculations naturally reveal the correct threshold noise strength D_c and the stationary solution exhibits *critical behavior*. Indeed, for D above D_c no nonzero a and b satisfy the Galerkin conditions, so that only the uniform solution P_0 satisfies the stationary GFPE (23) Physically, fluctuations (diffusion) overwhelm collisions mixing the limiting long-time distribution into an isotropic state. For D below D_c, however, the isotropic solution P_0 is modified by a non-isotropic correction that on approach to the critical diffusion strength satisfies the following asymptotic condition:

$$a = A(D_c - D)^{1/2} + A_1(D_c - D)^{3/2} + \mathcal{O}\left((D_c - D)^{5/2}\right), D \uparrow D_c. \tag{33}$$

It is easy to check that coefficient a is equal to the order parameter $|\overline{\tau}|$ and contains the leading perturbative term in $D_c - D$, with the critical amplitude $A \approx 0.706864$, and the critical exponent of $1/2$ from square root term, in agreement with [18]; the contribution from b is of higher order $\mathcal{O}(D_c - D)$.

Overall, FP_2 correctly predicts that the system undergoes a continuous phase transition (or a second-order transition) at the critical noise strength D_c. Furthermore, with D sufficiently close to D_c the above FP_2 has the correct qualitative behavior: $P^{(2)}$ it is positive and unimodal (i.e., with a single maximum at $\theta = 0$). For larger $D_c - D$, however, monotonicity breaks down as, eventually, does the positivity. Passing to FP_3 does not improve the situation as it loses monotonicity even sooner than FP_2 (i.e., for lower values of $D_c - D$; not shown). However, with IFP_3 (i.e., determining the 3-rd order coefficient c of $P^{(3)}$ using collocation at $\theta_1 = \pi$, rather than Galerkin):

$$c = \frac{\sqrt{K_1^2 - 4\pi K_2} - K_1}{2\pi}, \tag{34}$$

where

$$\begin{aligned} K_1 &= 1 + \tfrac{4}{3\pi} + 9D - \tfrac{24\pi a^2}{20D+5}, \\ K_2 &= \tfrac{b\left(3\pi^4 a^3 - 8\pi^2 a^2 (4D+1) + 3(4D+1)^2(\pi D + \pi - 4)\right)}{3\pi(4D+1)^2}. \end{aligned} \tag{35}$$

The improved Fourier polynomial approximation retains the correct qualitative behavior for D as low as $0.5D_c$, while the Fourier polynomial loses both monotonicity and positivity for θ away from 0.

As we are interested in accurate approximations to the stationary solution for the widest possible range of values of D, we further examine IFP_3 for $D = 0.4D$ and find it losing monotonicity. A higher order FP_4 is qualitatively wrong, similar to FP_2, and we discovered that improving that 4-th order Fourier polynomial with collocation at $\theta_1 = \pi$ is still inferior to IFP_3.

The situation can be improved by dropping the third order Galerkin condition and using two collocations at $\theta_1 = \pi$ and $\theta_2 = \pi/2$ to determine the two highest order coefficients of the 4-th order Fourier expansion. We do not present those results, however, and pass directly to IFP_5, which is determined, as is IFP_3 using a single collocation at $\theta_1 = \pi$ in addition to Galerkin conditions up to the 4-th order. The IFP_5 maintains correct qualitative properties for a wide range of D values; it further compares very favorably in quantitative terms to the numerical solution obtained in [5]. In summary, the presented method of (improved) Fourier polynomial approximations constitutes a rather flexible and effective tool for investigation of the stationary GFPE solutions.

3.2 Model 1: Stationary States with Self-Quenching Noise

Passing to Model 1, we consider first two cases $\alpha = 1$ and $\alpha = 2$, which are dramatically different in terms of their phase transition behavior. Taking $\alpha = 2$ first, the results remain qualitatively similar to Model 0 and, importantly, the critical diffusion D_c stays the same. The main difference between the two models reduces to a modified critical amplitude A. Indeed, applying Galerkin conditions we obtain

$$a = \sqrt{\frac{\sqrt{\det}}{24\pi D^2\kappa^2} - \frac{\pi}{6D^2\kappa^2} + \frac{5}{8D\kappa} - \frac{2}{\pi D\kappa} + \frac{1}{\kappa}},$$
$$\det = 3D^2\kappa\left(3(16-3\pi)^2\kappa - 64\pi^3\right) + 24\pi^2(16-5\pi)D\kappa + 16\pi^4,$$

and for the next FP coefficient,

$$b = -\frac{\pi b^2}{4Da^2\kappa - 4D - 1}. \tag{36}$$

Setting $\kappa = 1$, in the vicinity of the critical point the following asymptotic behavior holds for the order parameter

$$a \simeq A(D_c - D)^{1/2} + A_1(D_c - D)^{3/2} + \mathcal{O}\left((D_c - D)^{5/2}\right), \tag{37}$$

where the critical amplitude and the sub-amplitude are $A \approx 0.760696$, $A_1 \approx 1.19504$. When we set $\alpha = 1$, however, the effect of self-quenching becomes qualitatively different. Following the same computational route we obtain markedly different results.

The first difference to note is that the critical diffusion coefficient acquires a different value so that the new critical noise strength is greater than D_c:

$$D_c^* \simeq D_c + \frac{3\left(256 - 176\pi + 40\pi^2 - 3\pi^3\right)\kappa^2}{16\pi^4} + \mathcal{O}\left(\kappa^4\right), \kappa \downarrow 0, \tag{38}$$

where the κ^2 coefficient is strictly positive. Most importantly, as $D \uparrow D_c^*$ the order parameter *remains finite*:

$$a \simeq A^* + A(D_c^* - D)^{1/2} + O[D_c^* - D], \tag{39}$$

with

$$A^* = \frac{3(3\pi-16)\kappa\left(\pi\left(5\sqrt{3(\pi-4)\kappa^2+4\pi^2}+6\pi-32\right)-16\sqrt{3(\pi-4)\kappa^2+4\pi^2}\right)}{2\pi\left(3(16-5\pi)^2\kappa^2+64\pi^3\right)}, \tag{40}$$

$$A = \frac{\sqrt{\frac{3}{2\pi}}\sqrt{(3\pi-16)\sqrt{3(\pi-4)\kappa^2+4\pi^2}}}{2\left(\frac{48\kappa^2\left(\pi(16-3\pi)\sqrt{3(\pi-4)\kappa^2+4\pi^2}-10\pi^3+32\pi^2\right)^2}{\left(3(16-3\pi)^2\kappa^2-64\pi^3\right)^2}+\pi\right)}. \tag{41}$$

As observed earlier, with $k \downarrow 0$ Model 1 reduces to Model 0, and A approaches to the corresponding amplitude there. Thus, for $\kappa > 0, \alpha = 1$ Model 1 undergoes a *discontinuous* (or first-order) phase transition at D_c^*. The difference in behavior with $\alpha = 2$ can perhaps be explained by the fact that with $\alpha = 2$ the difference between $\overline{D}$ and D is an infinitesimal of higher order in $D_c - D$, so it does not affect the leading asymptotic behavior. With $\alpha = 1$ the influence of self-quenching is sufficiently strong to affect the leading-order behavior and render it discontinuous. We conjecture that there is a "critical" α_c lying between 1 and 2 at which the first-order transition is replaced by the second-order transition, but the value of this exponent is as yet unknown.

3.3 Elements of Kinetics

The behavior near the critical point, especially for Model 1, can be further elucidated by investigating its dynamic properties. To this end we extend the classical Landau-Hopf kinetic theory as follows. The time evolution of the slowly-varying part of the average orientation in the vicinity of the critical point is governed by the following equation [1, 13, 15]:

$$\frac{d\overline{\tau}}{dt} = (D_c - D)\overline{\tau} - \beta|\overline{\tau}|^2\overline{\tau}, \quad \beta > 0, \tag{42}$$

which is consistent with the stationary solution in the long-time limit: $\overline{\tau} \to \sqrt{\frac{D_c - D}{\beta}}$ as $t \uparrow \infty$.

Comparing with the previously derived expressions for the same quantity, we conclude that $\beta = \sqrt{\frac{2}{A}}$, where A stands for the critical amplitude for Model 0 and for Model 1 ($\alpha = 2$). This is basically the "spectral" approach of [1] in reverse: here we used our stationary solution to reconstruct the parameters of the equation describing the *relaxation* of the order parameter to its stationary value in the vicinity of the critical point.

Equation (42) can be solved exactly (see e.g., [21]). In particular, for $D_c < D$ we have $\overline{\tau}(t) \sim e^{-t(D-D_c)} \to 0$ with the characteristic relaxation time $(D - D_c)^{-1}$.

A different solution is obtained for $D_c = D$, namely $\overline{\tau}(t) \sim t^{-1/2} \rightarrow 0$, demonstrating the disappearance of a timescale dependence and an "infinite" characteristic relaxation time.

In order to characterize the "slope" of the order parameter time dependence in the general situation, let us introduce an *effective* characteristic relaxation time t_{eff}:

$$t_{eff}(t) = \frac{\overline{\tau}(\infty) - \overline{\tau}(t)}{\frac{d\overline{\tau}}{dt}}.$$

Thus, t_{eff} is itself a function of time and can be interpreted as the time necessary to reach the steady-state, if relaxing at the constant rate $\frac{d\overline{\tau}}{dt}$. As $t \rightarrow \infty$ this definition naturally recovers the constant characteristic relaxation time T for the aforementioned systems characterized by an asymptotically exponential decay. This is also the case for linearized Eq. (42), where we obtain $t_{eff}(\infty) = (D_c - D)^{-1}$, as expected. The divergence of the relaxation time of the order parameter, as the critical point is approached, is known as the *critical slowing down*.

For Model 1 ($\alpha = 2$) below the critical point, it turns out that a decrease in the non-linearity strength β (see Eq. (42)) does not change the form of $t_{eff}(\infty)$. Indeed, a decrease in β only increases the value of $t_{eff}(t)$ for finite times, leaving the long-time behavior of $t_{eff}(t)$ unaltered. Thus, the dynamics of Model 1 ($\alpha = 2$) are only marginally different from Model 0.

In order to capture the dynamics of Model 1 ($\alpha = 1$) with its discontinuous character of the phase transition, we need to introduce an additional term into the relaxation dynamics. We are going to look for the simplest polynomial extension to the right hand side of the (42) that in the limit of $t \uparrow \infty$ recovers the correct stationary state for Model 1 ($\alpha = 1$) as described above. The main properties that must be captured by the long-time solution are the positive shift of the critical diffusion strength $D_c^* - D_c > 0$, the finite value A of the order parameter at new critical diffusion, and the square-root form of the next-order dependence of a on the distance to criticality $D_c^* - D$. Assuming the kinetic ansatz

$$\frac{d\overline{\tau}}{dt} = (D_c - D)\overline{\tau} + \gamma|\overline{\tau}|\overline{\tau} - \delta|\overline{\tau}|^2\overline{\tau}, \quad \gamma > 0, \quad \delta > 0, \tag{43}$$

which generalizes (42), in the long-time limit we obtain a stationary state that has the correct form asymptotic behavior—a finite value at $D = D_c^*$ for suitable γ and δ:

$$\overline{\tau} = \frac{\sqrt{4\delta(D_c - D) + \gamma^2} + \gamma}{2\delta}. \tag{44}$$

By analogy with the simple kinetic model (42),the required parameter values are readily obtained by requiring that the steady-state match two known conditions: the value of critical point D_c^* and the positive stationary value of the order parameter A^* at $D = D_c^*$. The resulting equations are $D_c^* = \frac{\gamma^2}{4\delta} + D_c$, $\quad A^* = \frac{\gamma}{\delta}$. Thus, we have expressions relating the macroscopic observable parameters D_c^* and A^*, and the intrinsic mesoscopic kinetic parameters γ and δ: $\gamma = \frac{4(D_c^* - D_c)}{A^*}$, $\delta = \frac{4(D_c^* - D_c)}{A^{*2}}$. The corresponding relaxation time can be obtained explicitly:

$$t_{eff}^*(\infty) = \frac{2\delta}{\gamma\left(\sqrt{4\delta(D_c - D) + \gamma^2} + \gamma\right) + 4\delta(D_c - D)}. \tag{45}$$

Most importantly, $t_{eff}(\infty)$ remains *finite* at both points D_c and D_c^*. Moreover, as $D \uparrow D_c^*$ one finds

$$t_{eff}^*(\infty) \simeq \frac{\delta^{1/2}}{\gamma}(D_c^* - D)^{-1/2},$$

an inverse square root dependence for the discontinuous transition, which should be contrasted with the inverse linear dependence in the case of a continuous transition. Thus, compared with the basic Model 0 we have a dramatic speed-up of the relaxation to the stationary state because of the assumed self-quenching behavior. On the other hand, when $D \downarrow D_c^*$, we find an inverse linear dependence of the relaxation time $t_{eff}^*(\infty) \sim (D - D_c^*)^{-1}$ (see also [10]).

As was observed in [17], for groups of living active organisms near criticality (birds, fish, bacteria), the critical slowing down can play a detrimental role as it diminishes their ability to quickly relax to a new equilibrium, if required by the environment (e.g., the sudden appearance of a predator). At the same time, being "critical" dramatically enhances the group's *susceptibility* to external stimuli. Indeed, susceptibility denotes the response of the order parameter to an external stimulus h (for particles that are biological organisms the stimulus can be predators, drugs, source of nutrition), and can be quantified as the derivative $\frac{d\tau}{dh}$, which typically diverges at a second-order critical point as $h^{-2/3}$ [17]. Thus, at a continuous critical point cooperating particles can quickly determine the appropriate response (high susceptibility), but are slow to execute it (critical slowing down).

The self-quenching noise has the effect of inducing a high susceptibility at $D = D_c$, where it is still divergent, albeit at a weaker rate as $h^{-1/2}$. At the same time, at the nearby discontinuous transition point D_c^* the new equilibrium can be reached in finite time. Therefore, for D between D_c and D^* or nearby, high susceptibility to a stimulus coexists with a rapid relaxation to the new equilibrium determined by the changed environment. This behavior can be regarded as an evolutionary advantage that may have been employed, for example, by bacterial colonies for their survival. In particular, in order to fight bacteria more effectively one might have to turn off the effect of self-quenching noise, a peculiar mechanism allowing them to detect threats fast and react quickly.

In general, Model 1 exhibits a qualitatively different range of critical behavior, which depends on the value of the quenching exponent α. When $\alpha = 2$ Model 1 remains qualitatively similar to Model 0, although the self-quenching changes the critical amplitudes. For a fixed level of noise D this leads to a relative enhancement of the relaxation time compared with Model 0, but the transition remains continuous. As in all continuous transitions the relaxation time diverges at the critical point, resulting in the critical slowing down of the approach to the stationary state.

The proposed direct analytic approach to the solution of GFPE is sensitive enough to distinguish between this continuous transition and the discontinuous transition, which appears when the quenching exponent is $\alpha = 1$. The main result is the discovery of a discontinuous, abrupt transition from an ordered to a disordered state in this case. It is a macroscopic effect caused by a mesoscopic self-quenching noise. Here the relaxation time remains finite at the critical point, therefore the effect of self-quenching is to strongly suppress the critical slowing down and improve the reaction time to external stimuli. Nonetheless, in a peculiar way the system retains a characteristic of the continuous transition—a divergent sensitivity to external perturbations, quantified by the $h^{-1/2}$ divergence of the response to a stimulus of magnitude h. The physical significance of self-quenching noise, therefore, may be in the dual advantage it imparts to biological colonies: high sensitivity to a changing environment combined with fast execution of transition to a new, more desirable state. As first-order transitions are generally observed only in spatially-inhomogeneous particle systems, our translationally-invariant models must be interpreted as capturing the correct spatial averaging of the mechanisms of the phase transition behavior.

References

1. I.S. Aranson, L.S. Tsimring, Theory of self-assembly of microtubules and motors. Phys. Rev. E **74**, 031915, 1–15 (2006)
2. I.S. Aranson, A. Sokolov, J. Kessler, R. Goldstein, Model for dynamical coherence in thin films of self-propelled microorganisms. Phys. Rev. E **75**, 040901, 1–4 (2007)
3. G.K. Batchelor, J.T. Green, The determination of the bulk stress in a suspension of spherical particles to order c^2. J. Fluid Mech. **56**, 401–427 (1972)
4. C.W.J. Beenakker, The effective viscosity of a concentrated suspension (and its relation to diffusion). Physica A **128**, 48–81 (1984)
5. E. Ben-Naim, I. Krapivsky, Alignment of rods and partition of integers. Phys. Rev. E **73**, 031109, 1–7 (2006)
6. D.J. Bergman, Exact relations between critical exponents for elastic stiffness and electrical conductivity of two-dimensional percolating networks. Phys. Rev. E **65**, 026124-1–026124-7 (2002)
7. J. Bicerano, J.F. Douglas, D.A. Brune, Model for the viscosity of particle dispersions. Polym. Rev. **39**, 561–642 (1999)
8. B. Cichocki, M.L. Ekiel-Jezewska, E. Wajnryb, Three-particle contribution to effective viscosity of hard-sphere suspensions. J. Chem. Phys. **119**, 606–619 (2003)
9. R. Czapla, W. Nawalaniec, V. Mityushev, Effective conductivity of random two-dimensional composites with circular non-overlapping inclusions. Comput. Mater. Sci. **63**, 118–126 (2012)
10. M. Doi, S.F. Edwards, *The Theory of Polymer Dynamics* (Oxford University Press, Oxford, 1986)
11. K. Drescher, J. Dunkel, L. Cisneros, S. Ganguly, R. Goldstein, Fluid dynamics and noise in bacterial cell-cell and cell-surface scattering. Proc. Natl. Acad. Sci. U. S. A. **108**(27), 10940–10945 (2011)

12. S. Gluzman, V.I. Yukalov, Unified approach to crossover phenomena. Phys. Rev. E **58**, 4197–4209 (1998)
13. E. Hopf, A mathematical example displaying the features of turbulence. Commun. Pure Appl. Math. **1**, 303–322 (1948)
14. J.B. Keller, A theorem on the conductivity of a composite medium. J. Math. Phys. **5**, 548–549 (1964)
15. L.D. Landau, On the problem of turbulence. Dokl. Akad. Nauk SSSR **44**, 339–342 (1944, in Russian)
16. G.W. Milton, *The Theory of Composites*. Cambridge Monographs on Applied and Computational Mathematics, vol. 6 (Cambridge University Press, Cambridge, 2002)
17. T. Mora, W.J. Bialek, Are biological systems poised at criticality? J. Stat. Phys. **144**, 268–302 (2011)
18. F. Peruani, A. Deutch, M. Bar, A mean-field theory for self-propelled particles interacting by velocity alignment mechanisms. Eur. Phys. J. Spec. Top. **157**, 111–122 (2008)
19. P. Romanczuk, M. Bar, W. Ebeling, B. Lindner, L. Schimansky-Geier, Active Brownian particles. Eur. Phys. J. Spec. Top. **202**, 1–162 (2012)
20. A. Sokolov, I.S. Aranson, J. Kessler, R. Goldstein, Concentration dependence of the collective dynamics of swimming bacteria. Phys. Rev. Lett. **98**, 158102 (2007)
21. H.E. Stanley, *Introduction to Phase Transitions and Critical Phenomena* (Oxford University Press, Oxford, 1971)
22. D. Stauffer, A. Coniglio, M. Adam, Gelation and critical phenomena, in *Polymer Networks*, ed. by K. Dušek. Advances in Polymer Science, vol. 44 (Springer, Berlin, Heidelberg, 1982) pp. 103–158
23. S. Torquato, *Random Heterogeneous Materials: Microstructure and Macroscopic Properties* (Springer, New York, 2002)
24. E. Wajnryb, J.S. Dahler, The Newtonian viscosity of a moderately dense suspension. Adv. Chem. Phys. **102**, 193–313 (1997)
25. V.I. Yukalov, S. Gluzman, Critical indices as limits of control functions. Phys. Rev. Lett. **79**, 333–336 (1997)
26. E.P. Yukalova, V.I. Yukalov, S. Gluzman, Self-similar factor approximants for evolution equations and boundary-value problems. Ann. Phys. **323**, 3074–3090 (2008)

Mixed Problem for Laplace's Equation in an Arbitrary Circular Multiply Connected Domain

Vladimir Mityushev

Abstract Mixed boundary value problems for the two-dimensional Laplace's equation in a domain D are reduced to the Riemann-Hilbert problem Re $\overline{\lambda(t)}\psi(t) = 0$, $t \in \partial D$, with a given Hölder continuous function $\lambda(t)$ on ∂D except at a finite number of points where a one-sided discontinuity is admitted. The celebrated Keldysh-Sedov formulae were used to solve such a problem for a simply connected domain. In this paper, a method of functional equations is developed to mixed problems for multiply connected domains. For definiteness, we discuss a problem having applications in composites with a discontinuous coefficient $\lambda(t)$ on one of the boundary components. It is assumed that the domain D is a canonical domain, the lower half-plane with circular holes. A constructive iterative algorithm to obtain an approximate solution in analytical form is developed in the form of an expansion in the radius of the holes.

Keywords Mixed boundary value problem • Keldysh-Sedov formulae • Riemann-Hilbert problem • Multiply connected domain • Iterative functional equation

Mathematics Subject Classification (2010) 30E25

1 Introduction and Statement of the Problem

Mixed boundary value problems for the two-dimensional Laplace equation with the Dirichlet and Neumann boundary conditions have applications in mathematical physics when a potential is known on a part of the boundary L'' (the Dirichlet condition) and the normal flux on the rest part L' (the Neumann condition). The

---

V. Mityushev (✉)
Institute of Computer Sciences, Pedagogical University, ul. Podchorazych 2, Krakow 30-084, Poland
e-mail: mityu@up.krakow.pl

P. Drygaś, S. Rogosin (eds.), *Modern Problems in Applied Analysis*, Trends in Mathematics, https://doi.org/10.1007/978-3-319-72640-3_10

celebrated Keldysh-Sedov formulae [5] established in 1937 solve an arbitrary mixed boundary value problem for a half-plane in closed form. If we know a conformal mapping of the half-plane onto a domain D, we can solve an arbitrary mixed boundary value problem for this domain. It is worth noting that the Keldysh-Sedov formulae hold for any finite number of points $\mathcal{X}$ where L' and L'' meet. Various particular mixed problems were discussed in literature, for instance for a rectangle by means of Fourier series. Such problems can be easily solved in terms of the Weierstrass elliptic function $\wp(z)$ which conformally maps a rectangle onto the upper half-plane.

A mixed problem can be considered as a particular case of the Riemann-Hilbert problem [4, 12]

$$\operatorname{Re} \overline{\lambda(t)}\psi(t) = f(t), \quad t \in \partial D\backslash\mathcal{X}, \tag{1.1}$$

where the coefficient $\lambda(t)$ takes the value 1 on L'' and $i = \sqrt{-1}$ on L'. The finite set of points $\mathcal{X}$ divides the segments L' and L''. The function $f(t)$ is a given piece-wise Hölder continuous function on ∂D. Therefore, the Keldysh-Sedov formulae can be considered as exact formulae for solution to the special Riemann-Hilbert problem (1.1). The general Riemann-Hilbert problem for an arbitrary simply connected domain D, i.e. the problem (1.1) with arbitrary Hölder continuous function $\lambda(t)$, was solved in closed form in [4] up to a conformal mapping of D onto a canonical domain, for instance onto a half-plane. The problem was also solved for functions $\lambda(t)$ admitting a finite number of discontinuous points $\mathcal{X}$ where one-sided limits of $\lambda(t)$ exist [4].

The Riemann-Hilbert problem (1.1) for an arbitrary multiply connected domain D was solved in analytical form in [6] and systematically presented in [12] with some modifications [9]. Here, the term "analytic form" is used in accordance with the lines [4] and [12]. First, an exact solution for an arbitrary *circular* multiply connected domain $\mathbb{D}$ is obtained in terms of the Poincaré series. We say that an exact solution is obtained for an arbitrary multiply connected domain D if we know a conformal mapping of D onto a circular multiply connected domain $\mathbb{D}$. It was assumed in [6, 9, 12] that $\lambda(t)$ satisfies the Hölder condition on ∂D, i.e. $\mathcal{X} = \emptyset$. An extended historical survey can be found in [8, 10]. The concluding Sect. 5 of the present paper contains explanations concerning the papers [1, 15, 16].

In the present paper, we extend the results [6, 9, 12] to piecewise Hölder continuous coefficients $\lambda(t)$. For definiteness, we discuss a homogeneous mixed problem ($f(t) \equiv 0$) having applications to composites with a discontinuous coefficient $\lambda(t)$ on one of the components of ∂D. It is assumed that the domain D is a canonical domain, the lower half-plane with circular holes.

We proceed to precisely state the boundary value problem. Consider n mutually disjoint disks $|z - a_k| \leq r_k$ ($k = 1, 2, \ldots, n$) in the lower half-plane $\operatorname{Im} z < 0$ as displayed in Fig. 1. Let D denote the complement of the disks $|z - a_k| \leq r_k$ to the half-plane $\operatorname{Im} z < 0$. Let the x-axis be divided by finite points $x_1', x_1'', \cdots, x_p', x_p''$ onto $2p$ segments. Let $L' = \cup_{j=1}^{p}(x_j', x_j'')$ and L'' be the complement of the real axis

Fig. 1 Lower half-plane with circular inclusions

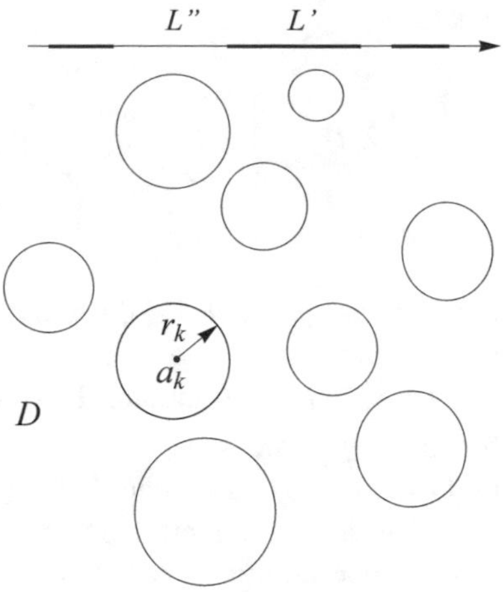

to the closure of L'. It is assumed that infinity belongs to L''. The problem is to find a function $\psi(z)$ analytic in D and continuous in its closure except the points x_j', x_j'' where it can have locally integrable (weak) singularities with the boundary conditions

$$\begin{aligned} \operatorname{Im} \psi(x) &= 0, \quad x \in L', \\ \operatorname{Re} \psi(x) &= 0, \quad x \in L'', \end{aligned} \tag{1.2}$$

$$\operatorname{Im} (t - a_k)\psi(t) = 0, \quad |t - a_k| = r_k \; k = 1, 2, \ldots, n. \tag{1.3}$$

The function $\psi(z)$ vanishes at infinity.

Introduce the set of points $\mathcal{X} = \cup_{j=1}^{p}(\{x_j'\} \cup \{x_j''\})$. The problem (1.2)–(1.3) is a Riemann-Hilbert problem for the domain D which can be written in the form [12]

$$\operatorname{Re} \overline{\lambda(t)}\psi(t) = 0, \quad t \in \partial D \backslash \mathcal{X}, \tag{1.4}$$

where

$$\lambda(t) = \begin{cases} -i, & t \in L', \\ 1, & t \in L'', \\ -\dfrac{ir_k}{t - a_k}, & |t - a_k| = r_k \; k = 1, 2, \ldots, n. \end{cases} \tag{1.5}$$

The orientation of ∂D is opposite to the orientation of the real axis and to the counter clockwise orientation of the circles $|t - a_k| = r_k$. The modulus of $\lambda(t)$ is equal to unity. The condition (1.4) can be also written in the form

$$\psi(t) = -\frac{\lambda(t)}{\overline{\lambda(t)}} \overline{\psi(t)}, \quad t \in \partial D \backslash \mathcal{X}, \tag{1.6}$$

where

$$\frac{\lambda(t)}{\overline{\lambda(t)}} = \begin{cases} -1, & t \in L', \\ 1, & t \in L'', \\ -\left(\dfrac{r_k}{t-a_k}\right)^2, & |t-a_k| = r_k,\ k = 1,2,\ldots,n. \end{cases} \tag{1.7}$$

In accordance with [4, 12] the winding number (index) of the problem (1.6) is equal to $\kappa = ind_{\partial D}\lambda(t) = p + n$. Here, every pair of discontinuity points on the real axis (x_j', x_j'') gives the amount 1 to the winding number. The winding number of the function $-\frac{ir_k}{t-a_k}$ along the circle $|t-a_k| = r_k$ is equal to (-1) what gives the amount $(+1)$ to the winding number of the problem.

The Riemann-Hilbert problem (1.2)–(1.3) has important applications to composites. For definiteness, let us consider the thermal conduction described by the temperature distribution $u(x,y) \equiv u(z)$. The function $u(z)$ satisfies Laplace's equation in the domain D and continuously differentiable in its closure except in $\mathcal{X}$ where it is bounded. The gradient ∇u is continuous in the closure of D except in $\mathcal{X}$ where it can have integrable singularities. Let the segments L' be isolators. Then, the heat flux normal to the real axis vanishes on L'

$$\frac{\partial u}{\partial y} = 0, \quad \text{on } L'. \tag{1.8}$$

Let the temperature $u(z)$ attain a constant value d_j at each component L_j'' of L'' and c_k at every circle $|t-a_k| = r_k$:

$$u(x,0) = d_j, \quad x \in L_j'', \quad j = 1,2,\ldots,p, \tag{1.9}$$

$$u(t) = c_k, \ |t-a_k| = r_k, \quad k = 1,2,\ldots,n. \tag{1.10}$$

Introduce the complex flux analytic in D and having the boundary properties described above

$$\psi(z) = \frac{\partial u}{\partial x} - i\frac{\partial u}{\partial y}. \tag{1.11}$$

Then, the boundary condition (1.8) takes the form of the first condition (1.2). Differentiate (1.9)

$$\frac{\partial u}{\partial x}(x,0) = 0, \quad x \in L''. \tag{1.12}$$

This equation is equivalent to the second condition (1.2).

Fix a circle $|t-a_k| = r_k$ and consider the unit tangent $\mathbf{s} = (-n_2, n_1)$ and normal $\mathbf{n} = (n_1, n_2)$ vectors to it. Continuous differentiability of $u(x,y)$ on $|t-a_k| = r_k$

implies existence of the directional derivative

$$\frac{\partial u}{\partial \mathbf{s}} = -n_2 \frac{\partial u}{\partial x} + n_1 \frac{\partial u}{\partial y}. \tag{1.13}$$

Equation (1.10) can be differentiated in the tangent direction **s**

$$\frac{\partial u}{\partial \mathbf{s}} = 0, \quad |t - a_k| = r_k. \tag{1.14}$$

Let $n = n_1 + in_2$ be the vector $\mathbf{n} = (n_1, n_2)$ written in the complex form. Then, the tangent derivative (1.13) can be written as follows

$$\frac{\partial u}{\partial \mathbf{s}} = -\mathrm{Im}\,[n(t)\ \psi(t)], \quad |t - a_k| = r_k. \tag{1.15}$$

The normal vector to the circle $|t - a_k| = r_k$ can be written in the form $n(t) = \frac{t-a_k}{r_k}$. Therefore, the boundary condition (1.14) becomes (1.3). Thus, the problem (1.2)–(1.3) describes the steady heat flux in D with insulation L' and given constant temperatures at the components of L'' and at $|t - a_k| = r_k$ $(k = 1, 2, \ldots, n)$. These constants can be found by integration of $\psi(z)$ after solution to the problem (1.2)–(1.3) since the general solution $\psi(z)$ depends on $p + n$ real constants [4, 12].

2 Factorization

2.1 Factorization on the Real Axis

We now describe the branch of the square root

$$R(z) = \sqrt{\prod_{j=1}^{p} (z - x_j')(z - x_j'')}. \tag{2.1}$$

Let the function $R(z)$ be analytic in the complex plane except the slits L'. The function $R(z)$ is continuous in the considered domains and has a pole at infinity, $R(z) \sim z^p$, as $z \to \infty$. It is assumed that $R(x)$ is positive for $x \in L''$. The limit values of $R(z)$ are pure imaginary on L' and have the form

$$\begin{aligned} R(x + i0) = R^+(x) &= i\sqrt{-\prod_{j=1}^{p} (x - x_j')(x - x_j'')}, \\ R(x - i0) = R^-(x) &= -i\sqrt{-\prod_{j=1}^{p} (x - x_j')(x - x_j'')},\ x \in L'. \end{aligned} \tag{2.2}$$

Let P_j $(j = 0, 1, \cdots, p-1)$ be real constants. Following [4, p. 467] introduce the function

$$\psi_0(z) = i\,\frac{P_0 + P_1 z + \cdots + P_{p-1} z^{p-1}}{R(z)}. \tag{2.3}$$

It is analytic in the lower half-plane and continuous in its closure except the points $\mathcal{X}$ where it has weak singularities. The polynomial

$$\omega_0(z) = i(P_0 + P_1 z + \cdots + P_{p-1} z^{p-1}) \tag{2.4}$$

solves the boundary value problem

$$\mathrm{Re}\ \omega_0(x) = 0,\ x \in \mathbb{R} \tag{2.5}$$

in the class of function analytic in the lower half-plane and continuous in its closure except infinity where $\omega_0(z)$ has a pole of order $p-1$. We have $\omega_0(z) = R(z)\psi_0(z)$. Hence, the function $\psi_0(z)$ solves the boundary value problem in the considered class [4]

$$\mathrm{Im}\ \psi_0(x) = 0,\ x \in L', \quad \mathrm{Re}\ \psi_0(x) = 0,\ x \in L''. \tag{2.6}$$

The function $R(z)$ is called the factorization function for the problem (2.6) [4]. It satisfies the boundary condition

$$\mathrm{Re}\ R^-(x) = 0,\ x \in L', \quad \mathrm{Im}\ R^-(x) = 0,\ x \in L''. \tag{2.7}$$

2.2 General Factorization

We introduce the factorization function $X(z)$ for the problem (1.6). It is analytic in the domain D and continuous in its closure except at infinity where it has a pole of order $2n + p$, more precisely $\lim_{z\to\infty} z^{-(2n+p)} X(z) = 1$. The factorization function $X(z)$ does not vanish in the closure of D except perhaps the set $\mathcal{X}$ and satisfies the boundary condition [12]

$$\frac{X(t)}{\overline{X(t)}} = \frac{\overline{\nu_k \lambda(t)}}{\nu_k \lambda(t)},\quad |t - a_k| = r_k,\ k = 1, 2, \cdots, n, \tag{2.8}$$

$$\frac{X(x)}{\overline{X(x)}} = \frac{\overline{\lambda(x)}}{\lambda(x)},\quad x \in \mathbb{R}. \tag{2.9}$$

The inverse expression to $\frac{\overline{\lambda(t)}}{\lambda(t)}$ is given by (1.7). The constants ν_k have to be found during solution to the problem (2.8)–(2.9). One can assume that $|\nu_k| = 1$.

The problem (2.8)–(2.9) can be reduced to the following multiplicative $\mathbb{R}$-linear problem[1]

$$X(t) = X_k(t)\overline{X_k(t)}\ i\ \overline{\nu_k}\ \frac{t-a_k}{r_k}, \quad |t-a_k| = r_k,\ k = 1, 2, \cdots, n, \tag{2.10}$$

$$X(x) = X_0(x)\overline{X_0(x)}\ R^-(x), \quad x \in \mathbb{R}. \tag{2.11}$$

The functions $X_k(z)$ and $X_0(z)$ are analytic in $|z - a_k| < r_k$ and in $\mathrm{Im}\ z > 0$, respectively, and continuous in the closures of the considered domains. It is assumed that $X_k(z)$, $X_0(z)$ and $X(z)$ do not vanish in $|z - a_k| \leq r_k$, $\mathrm{Im}\ z \geq 0$ and $D \cup \partial D$, respectively, except perhaps the set $\mathcal{X}$.

Moreover, the normalization $X_0(z) \sim z^n$, as $z \to \infty$, can be taken without loss of generality. Another normalization $X_k(a_k) > 0$ follows from the invariance of the change $X_k(z) \longmapsto c_k X_k(z)$ for $|c_k| = 1$ in (2.10). One can see that the condition (2.11) is fulfilled at infinity since the left and right sides have the same polynomial behavior as $z \to \infty$.

The problems (2.8)–(2.9) and (2.10)–(2.11) are equivalent. More precisely, we have

Theorem 2.1

i) The problem (2.8)–(2.9) *is solvable if and only if* (2.10)–(2.11) *is solvable.*
ii) If (2.10)–(2.11) *has a solution* $X(z)$*, it is a solution of* (2.8)–(2.9).
iii) Conversely, if (2.8)–(2.9) *has a solution* $X(z)$*, it is a solution of* (2.10)–(2.11) *in D. Solutions* $X_k(z)$ *in* D_k *and* $X_0(z)$ *in* $\mathrm{Im}\ z > 0$ *of* (2.10)–(2.11) *can be found from simple problems for the simply connected domains* D_k *and* $\mathrm{Im}\ z > 0$*. For instance, if the positive function* $\frac{X(x)}{R^-(x)}$ *is given, the function* $X_0(z)$ *can be found from the modulus boundary value problem*

$$|X_0(x)| = \sqrt{\frac{X(x)}{R^-(x)}}, \quad x \in \mathbb{R}. \tag{2.12}$$

The problem (2.12) has such a solution that $X_0(z) \sim z^n$*, as* $z \to \infty$*. Moreover,* $X_0(z)$ *does not vanish in* $\mathrm{Im}\ z \geq 0$ *perhaps except* $\mathcal{X}$*.*

Proof The proof follows the lines of [8, Lemma 1] where analogous assertion was proved for the $\mathbb{R}$-linear problem in the standard additive form. We should add only that following [12] the modulus problem (2.12) is reduced to the additive problem

[1]The standard form for the $\mathbb{R}$-linear problem is additive [9, 12], e.g. $\phi(t) = \phi_k(t) + \overline{\phi_k(t)} + c(t)$. In this case $X(t) = \exp\phi(t)$, $X_k(t) = \exp\phi_k(t)$.

with respect to $\ln X_0(z)$ (Schwarz's problem for the upper half-plane)

$$\text{Re}\,[\ln X_0(x)] = \ln\sqrt{\frac{X(x)}{R^-(x)}}, \quad x \in \mathbb{R}. \tag{2.13}$$

The logarithm is correctly defined since $X_0(z)$ does not vanish in D. □

Let $z^*_{(k)}$ denote the inversion (reflection) of a point z with respect to the circle $|t - a_k| = r_k$

$$z^*_{(k)} = \frac{r_k^2}{\overline{z - a_k}} + a_k, \tag{2.14}$$

i.e., the points z and $z^*_{(k)}$ are symmetric with respect to $|t - a_k| = r_k$. If a function $f(z)$ is analytic in $|z - a_k| < r_k$, then the function $\overline{f(z^*_{(k)})}$ is analytic in $|z - a_k| > r_k$ [12].

Introduce the function piecewise analytic in the considered domains

$$\Omega(z) = \begin{cases} \dfrac{X_k(z)}{R(z)} \left[\overline{X_0(\overline{z})} \prod\limits_{m \neq k} (z - a_m)\overline{X_m(z^*_{(m)})}\right]^{-1} \dfrac{i\,\overline{v_k}}{r_k}, & |z - a_k| \le r_k, \\ & k = 1, 2, \cdots, n, \\ X_0(z) \left[\prod\limits_{m=1}^{n} (z - a_m)\overline{X_m(z^*_{(m)})}\right]^{-1}, & \text{Im}\, z \ge 0, \\ \dfrac{X(z)}{R(z)} \left[\overline{X_0(\overline{z})} \prod\limits_{m=1}^{n} (z - a_m)\overline{X_m(z^*_{(m)})}\right]^{-1}, & z \in D, \end{cases} \tag{2.15}$$

where m runs over $1, 2, \ldots, n$ except k in the product $\prod_{m \neq k}$. Using (2.10)–(2.11) one can see that $\Omega^+(t) = \Omega^-(t)$ on the circles $|t - a_k| = r_k$ and on the real line. This implies the analytic continuation of $\Omega(z)$ through the circles and the real axis. Then, application of the Liouville theorem implies that $\Omega(z)$ is a constant. Calculate this constant at infinity using (2.15)

$$\Omega(z) = \lim_{z \to \infty} X_0(z) \left[\prod_{m=1}^{n} (z - a_m)\overline{X_m(z^*_{(m)})}\right]^{-1} = \prod_{m=1}^{n} \frac{1}{\overline{X_m(a_m)}}, \tag{2.16}$$

where $X_m(a_m) > 0$. Using the definition of $\Omega(z)$ we arrive at the system of functional equations

$$X_k(z) = -\frac{i\,\nu_k r_k R(z)}{X_k(a_k)}\overline{X_0(\overline{z})}\prod_{m\neq k}(z-a_m)\frac{\overline{X_m(z^*_{(m)})}}{X_m(a_m)}, \tag{2.17}$$

$$|z-a_k| \leq r_k,\ k=1,2,\ldots,n,$$

$$X_0(z) = \prod_{m=1}^{n}(z-a_m)\frac{\overline{X_m(z^*_{(m)})}}{X_m(a_m)}, \quad \operatorname{Im} z \geq 0. \tag{2.18}$$

The function $X_0(z)$ can be eliminated from the system (2.17)–(2.18). It follows from (2.18) that

$$\overline{X_0(\overline{z})} = \prod_{m=1}^{n}(z-\overline{a_m})\frac{X_m(\overline{z}^*_{(m)})}{X_m(a_m)}, \quad \operatorname{Im} z < 0. \tag{2.19}$$

Substitution of (2.19) into (2.17) yields

$$X_k(z) = -\frac{i\,\nu_k r_k R(z)}{X_k(a_k)}\prod_{m=1}^{n}(z-\overline{a_m})\frac{X_m(\overline{z}^*_{(m)})}{X_m(a_m)}\prod_{m\neq k}(z-a_m)\frac{\overline{X_m(z^*_{(m)})}}{X_m(a_m)}, \tag{2.20}$$

$$|z-a_k| \leq r_k,\ k=1,2,\ldots,n.$$

After construction of the functions $X_k(z)$ the function $\overline{X_0(\overline{z})}$ is determined by (2.19). The factorization function $X(z)$ is found from the definition (2.15) of $\Omega(z)$ and (2.19)

$$X(z) = R(z)\prod_{m=1}^{n}(z-\overline{a_m})(z-a_m)\frac{X_m(\overline{z}^*_{(m)})\overline{X_m(z^*_{(m)})}}{X_m^2(a_m)}, \quad z \in D. \tag{2.21}$$

2.3 *Solution to Functional Equations*

We now proceed to write functional equations (2.20) in the additive form following [12]. Introduce the single-valued functions $\varphi_k(z) = \ln X_k(z)$ analytic in $|z-a_k| < r_k$ $(k = 1, 2, \ldots, n)$ and continuous in $|z-a_k| \leq r_k$. Then, (2.20) can be written in the form

$$\varphi_k(z) = \sum_{m=1}^{n}[\varphi_m(\overline{z}^*_{(m)}) - \varphi_m(a_m)] + \sum_{m\neq k}[\overline{\varphi_m(z^*_{(m)})} - \varphi_m(a_m)] + \tag{2.22}$$

$$F_k(z) + c_k, \quad |z-a_k| \leq r_k,\ k=1,2,\ldots,n,$$

where

$$F_k(z) = \sum_{m=1}^{n} \ln(z - \overline{a_m}) + \sum_{m \neq k} \ln(z - a_m) + \ln[r_k R(z)], \tag{2.23}$$

and

$$c_k = -\varphi_k(a_k) + \ln(-i\nu_k), \tag{2.24}$$

where Re $c_k = -\varphi_m(a_k)$ and Im $c_k = \arg(-i\nu_k)$ since $|\nu_k| = 1$. It follows from [12, Sec.4.3] that for given $F_k(z) + c_k$ the system of functional equations (2.22) has a unique solution. However, the constant c_k includes $\varphi_k(a_k)$ which yields solvability conditions for (2.22) [12, Sec.4.3]. Because of the linearity of (2.22) we can construct $\varphi_k(z)$ up to an additive constants [12, p. 151] by means of $F_k(z)$. More precisely, consider the modified system (2.22)

$$\widetilde{\varphi_k}(z) = \sum_{m=1}^{n} [\widetilde{\varphi_m}(\overline{z}^*_{(m)}) - \widetilde{\varphi_m}(a_m)] + \sum_{m \neq k} [\overline{\widetilde{\varphi_m}(z^*_{(m)})} - \overline{\widetilde{\varphi_m}(a_m)}] + \tag{2.25}$$
$$F_k(z), \quad |z - a_k| \leq r_k, \ k = 1, 2, \ldots, n.$$

Theorem 2.2 ([12]) *The system of functional equations* (2.25) *has a unique solution which can be found by uniformly convergent successive approximation.*

This theorem was proved in [12, Sec.4.3] in slightly different form as described below.

1. Instead of $\widetilde{\varphi_m}(a_m)$ the constants $\widetilde{\varphi_m}(w^*_{(m)})$ were used.
2. The functional equations can be written in the form used in [12]

$$\Phi_k(z) = \sum_{m \neq k} [\overline{\Phi_m(z^*_{(m)})} - \overline{\Phi_m(a_m)}] + F_k(z), \quad |z - a_k| \leq r_k, \ k = 1, 2, \ldots, 2n, \tag{2.26}$$

   where $\Phi_k(z) := \widetilde{\varphi_k}(z)$, $|z - a_k| \leq r_k$ $(k = 1, 2, \ldots, n)$ and $\Phi_k(z) := \overline{\widetilde{\varphi_{k-n}}(\overline{z})}$, $|z - a_k| \leq r_k$, $a_k := \overline{a}_{k-n}$ $(k = n+1, n+2, \ldots, 2n)$. The same designations are used for $F_k(z)$.
3. The functions $\Phi_k(z)$ were written in [12] in the form of the uniformly convergent series

$$\Phi_k(z) = F_k(z) + \sum_{m_1 \neq k} [\overline{F_{m_1}(z^*_{(m_1)})} - \overline{F_{m_1}(a_{m_1})}] + \tag{2.27}$$
$$\sum_{m_1 \neq k} \sum_{m_2 \neq m_1} [F_{m_2}(z^*_{(m_2, m_1)}) - F_{m_2}(a_{m_2})] +$$

$$\sum_{m_1 \neq k} \sum_{m_2 \neq m_1} \sum_{m_3 \neq m_2} [\overline{F_{m_1}(z^*_{((m_2,m_1,m_3))})} - \overline{F_{m_1}(a_{m_3})}] + \ldots,$$

$$|z - a_k| \leq r_k, \ k = 1, 2, \ldots, 2n,$$

where the composition of successive inversions with respect to the circles $k_1, k_2, \ldots, k_s$ are used

$$z^*_{(k_s k_{s-1} \ldots k_1)} := \left(z^*_{(k_{s-1} \ldots k_1)} \right)^*_{(k_s)}. \tag{2.28}$$

The series obtained from

$$\theta_2(z) = \sum_{k=1}^{n} \overline{\Phi_k(z^*_{(k)})} \tag{2.29}$$

by substitution of the series (2.27) is called the θ_2-Poincaré associated to the Schottky group [12]. This Schottky group consists of the compositions of inversions with respect to the circles $|t - a_k| = r_k$ $(k = 1, 2, \ldots, 2n)$ given by (2.28).

Now, substitute $z = a_k$ into (2.22) and take the imaginary part

$$\mathrm{Im}\, \{ \sum_{m=1}^{n} [\widetilde{\varphi_m}((\overline{a_k})^*_{(m)}) - \widetilde{\varphi_m}(a_m)] + \sum_{m \neq k} [\overline{\widetilde{\varphi_m}((a_k)^*_{(m)})} - \widetilde{\varphi_m}(a_m)] \} + \tag{2.30}$$

$$\mathrm{Im}\, F_k(a_k) + \arg(-i\nu_k) = 0, \quad k = 1, 2, \ldots, n.$$

These equations determine the constants ν_k (see the corresponding equations (4.4.19) from [12]).

It follows from (2.21) that the function $\ln X(z)$ is related to the functions $\varphi_k(z)$ by formulae

$$\ln X(z) = \sum_{m=1}^{n} \{ [\widetilde{\varphi_m}(\overline{z}^*_{(m)}) - \widetilde{\varphi_m}(a_m)] + [\overline{\widetilde{\varphi_m}(z^*_{(m)})} - \widetilde{\varphi_m}(a_m)] \} + \tag{2.31}$$

$$\ln R(z) + \sum_{m=1}^{n} \ln(z - a_m)(z - \overline{a_m}), \quad z \in D.$$

The representation of the function $\widetilde{\varphi_k}(z) = \Phi_k(z)$, $(k = 1, 2, \ldots, n)$ in the form of infinite series and substitution of these series into (2.21) express $\ln X(z)$ in terms of the uniformly convergent θ_2-Poincaré (2.27), (2.29). Though we do not write the final formula here, we may say that the factorization function can be constructed exactly in analytic form.

3 Solution to the Main Problem

We now proceed to solve the Riemann-Hilbert problem (1.6) by means of the factorization method [4]. Equations (2.8)–(2.9) can be considered as the representation of the coefficient $\frac{\lambda(t)}{\overline{\lambda(t)}}$ from (1.6) in the following form

$$\frac{\lambda(t)}{\overline{\lambda(t)}} = \frac{\overline{\nu_k}}{\nu_k}\frac{\overline{X(t)}}{X(t)}, \quad |t-a_k| = r_k, \ k = 1, 2, \cdots, n, \tag{3.1}$$

$$\frac{\lambda(x)}{\overline{\lambda(x)}} = \frac{\overline{X(x)}}{X(x)}, \quad x \in \mathbb{R}. \tag{3.2}$$

Introduce the function analytic in D

$$\omega(z) = X(z)\psi(z). \tag{3.3}$$

Using (2.8)–(2.9) we rewrite (1.6) as the Riemann-Hilbert problem with respect to $\omega(z)$

$$\text{Re } \nu_k\omega(t) = 0, \quad |t-a_k| = r_k, \ k = 1, 2, \cdots, n, \tag{3.4}$$

$$\text{Re } \omega(x) = 0, \quad x \in \mathbb{R}, \tag{3.5}$$

where ν_k are given constants for which $|\nu_k| = 1$. The function $\omega(z)$ is continuous in $D \backslash \mathcal{X}$ and almost bounded at the points $\mathcal{X}$ [4]. It has a pole at infinity of order $2n + p - 1$. The Riemann-Hilbert problem (3.4)–(3.5) with constant coefficients at each component of the circular boundary (a straight line is a circle in complex analysis) was solved in [12, Theorem 4.13] in analytical form in terms of the uniformly convergent Poincaré α-series [11] constructed analogously to (2.27), (2.29). Let $0 \leq \alpha_j < 2\pi$ $(j = 1, 2, \ldots, n)$. The series

$$\alpha_2(z) = \sum_{k=1}^{n} e^{i\alpha_k}\overline{\Phi_k(z_{(k)}^*)} \tag{3.6}$$

with

$$\begin{aligned}\Phi_k(z) = F_k(z) + \sum_{m_1 \neq k} e^{i\alpha_{m_1}}[\overline{F_{m_1}(z_{(m_1)}^*)} - \overline{F_{m_1}(a_{m_1})}] + \\ \sum_{m_1 \neq k}\sum_{m_2 \neq m_1} e^{i(\alpha_{m_1}-\alpha_{m_2})}[F_{m_2}(z_{(m_2,m_1)}^*) - F_{m_2}(a_{m_2})] + \\ \sum_{m_1 \neq k}\sum_{m_2 \neq m_1}\sum_{m_3 \neq m_2} e^{i(\alpha_{m_1}-\alpha_{m_2}+-\alpha_{m_3})}[\overline{F_{m_1}(z_{((m_2,m_1,m_3))}^*)} - \overline{F_{m_1}(a_{m_3})}] + \ldots, \\ |z-a_k| \leq r_k, \ k = 1, 2, \ldots, 2n,\end{aligned} \tag{3.7}$$

is called the Poincaré α-series.

It follows from [9] that the general solution of the problem (3.4)–(3.5) linearly depends on $n + p$ arbitrary real constants and has the following structure, e.g. (2.4)

$$\omega(z) = P_1\omega_1(z) + P_2\omega_2(z) + \cdots + P_{n+p}\omega_{n+p}(z). \tag{3.8}$$

It follows from (3.3) that

$$\psi(z) = \frac{\omega(z)}{X(z)}, \quad z \in D. \tag{3.9}$$

Formula (3.9) implies that $\psi(z)$ is analytic in D, continuous in its closure including infinity except in $\mathcal{X}$ where it can have weak singularities, e.g. (2.3). Therefore, the solution of the corresponding Riemann-Hilbert problem (3.9) can be exactly written as the ratio of the linear combination of the Poincaré α-series to the θ_2-Poincaré type product.

4 Approximation in Radius

In the present section, we develop a symbolic-numerical algorithm for the factorization function. For definiteness, the case of equal disks $|z-a_k| < r$ is considered. The method is based on the approximations in r^2. For simplicity, we discuss here only the main term, as $r \to 0$. Small r means that the ratios of the radius r to the distances between the centers a_k, a_m $(m \neq k)$ and to the distances to the real line Im a_k are sufficiently small. It is worth noting that the method of successive approximations applied to the functional equations (2.25), hence to Eqs. (2.20), yields an analytic dependence of $\frac{X_k(z)}{X_k(a_k)}$ on the parameter r^2 that justifies the proposed asymptotics. The iterative scheme presented below can be considered as a numerical estimation of the corresponding Poincaré products in radius. Such a construction yields simple approximate analytical formulae useful in applications for local concentrations of disks in the lower half-plane about 10–20% [13, 14].

Let $\frac{X_m(z)}{\alpha_m} = 1 + \beta_m(z - a_m) + \ldots$ and $\beta_m = \frac{X_k'(a_m)}{\alpha_m}$ where $\alpha_m = X_m(a_m) > 0$. We have

$$\frac{\overline{X_m(z^*_{(m)})}}{\overline{X_m(a_m)}} = 1 + \overline{\beta_m}\frac{r^2}{z - a_m} + O(r^4), \tag{4.1}$$

$$\frac{X_m(\overline{z}^*_{(m)})}{X_m(a_m)} = 1 + \beta_m\frac{r^2}{z - \overline{a_m}} + O(r^4), \tag{4.2}$$

Then, the functional equation (2.20) yields

$$X_k(z) = -r\frac{i\,\nu_k Q_k(z)}{\alpha_k}\left[1 + r^2\left(\sum_{m=1}^{n}\frac{\beta_m}{z-a_m} + \sum_{m\neq k}\frac{\overline{\beta_m}}{z-\overline{a_m}}\right) + \ldots\right], \tag{4.3}$$

$$|z-a_k| \leq r_k,\ k = 1,2,\ldots,n,$$

where

$$Q_k(z) = R(z)\prod_{m=1}^{n}(z-\overline{a_m})\prod_{m\neq k}(z-a_m)\,. \tag{4.4}$$

Substituting $z = a_k$ into (4.3) and multiplying the result by α_k we obtain

$$\alpha_k^2 = -ri\,\nu_k Q_k(a_k)\left[1 + r^2\left(\sum_{m=1}^{n}\frac{\beta_m}{a_k-a_m} + \sum_{m\neq k}\frac{\overline{\beta_m}}{a_k-\overline{a_m}}\right) + \ldots\right], \tag{4.5}$$

$$k = 1,2,\ldots,n.$$

Taking the leading term in (4.5) we obtain

$$\alpha_k^2 = -ri\,\nu_k Q_k(a_k) + O(r^3), \quad k = 1,2,\ldots,n. \tag{4.6}$$

Equation (4.6) determines the zero-th approximation of ν_k for which $|\nu_k| = 1$ and

$$-i\nu_k Q_k(a_k) > 0, \quad k = 1,2,\ldots,n. \tag{4.7}$$

When ν_k is determined, the approximation of α_k is found from (4.6)

$$\alpha_k = \sqrt{r}\sqrt{-i\,\nu_k Q(a_k)} + O(r^{5/2}), \quad k = 1,2,\ldots,n. \tag{4.8}$$

One can see that the leading term of α_k is of order $\sqrt{r}$, as $r \to 0$.

The iteration scheme can be proceeded to higher order terms in r^2 and the Taylor coefficients of $X_k(z)$ can be calculated. Every Taylor coefficient can be presented as a product of $\sqrt{r}$ by an analytic in r^2 function.

Formula (2.21) yields the approximations

$$X(z) = R(z)\prod_{m=1}^{n}(z-\overline{a_m})(z-a_m) + O(r^2), \quad z \in D. \tag{4.9}$$

Following formulae [12, p.167] we obtain up to $O(r^4)$

$$\omega(z) \approx -\left[\sum_{m=1}^{n}\overline{\nu_m}[\overline{\Phi(z_{(m)}^*)} - \overline{\Phi(a_m)}] - \sum_{m=1}^{n}\nu_m[\Phi(\overline{z}_{(m)}^*) - \Phi(a_m)]\right], \quad z \in D, \tag{4.10}$$

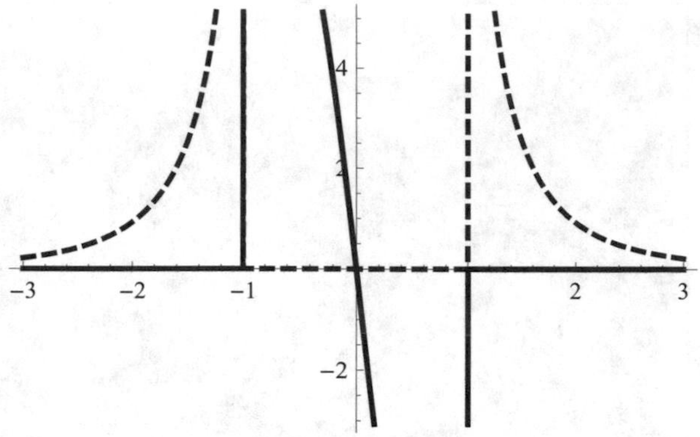

Fig. 2 Re $\psi(x)$ (solid line) and Im $\psi(x)$ (dashed line) calculated on the real axis by (4.12) with $\gamma = 10^3, l = 1, h = 1$

where $\Phi(z)$ is a polynomial of power $2n + p - 1$. One can see that the leading term in (4.10) is of order r^2

$$\omega(z) = -r^2 \sum_{m=1}^{n} \left[\overline{\nu_m \frac{\Phi'(a_m)}{z - a_m}} - \nu_m \frac{\Phi'(a_m)}{z - \overline{a_m}} \right] + O(r^4), \quad z \in D, \tag{4.11}$$

Therefore, the function $\psi(z)$ is calculated by the asymptotic formulae (3.9), (4.11) and (4.9).

Consider an example. Let $n = 1, p = 1, a_1 = -ih, x_1' = -l, x"_1 = l$ where h and l are positive numbers. Then, $R(z) = \sqrt{z^2 - l^2}$ and $Q_k(z) = (z - ih)\sqrt{z^2 - l^2}$. The conditions (4.7) give $\nu_1 = -i$ and $\Phi'(-ih) = i\gamma$ where γ is a real number. The function $\psi(z)$ can be calculated up to $O(r^4)$

$$\psi(z) \approx -\frac{2h\, i\gamma\, r^2}{(z^2 + h^2)^2 \sqrt{z^2 - l^2}}, \quad z \in D. \tag{4.12}$$

The results of computations are displayed in Figs. 2 and 3.

5 Discussion

In the present paper, solution to the mixed boundary value problem and the corresponding Riemann-Hilbert problem (3.9) is constructed. It can be exactly written as the ratio of the linear combination of the Poincaré α-series to the θ_2-Poincaré type product. One can read in literature that the above mixed boundary value problem had been solved and the corresponding Keldysh-Sedove formulae had been constructed [15, 16]. This requires explanations.

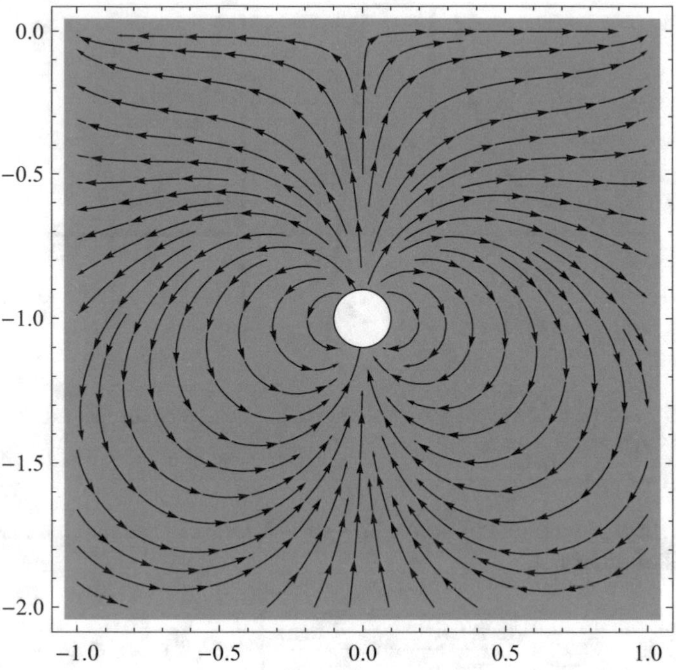

Fig. 3 Streamlines of the complex flux $\psi(z)$ for the same data as in Fig. 2

First, we discuss the separation condition [2, 3], a geometrical restriction on the circles $|t - a_k| = r_k$ $(k = 1, 2, \ldots, n)$. Roughly speaking the separation condition means that the circles have to be far way from each other. All the known exact formulae in the form of the *absolutely and uniformly* convergent θ_2-Poincaré series or equivalently in the form of the Schottky-Klein prime function are applied under such a separation condition that restricts application of these formulae [2, 3]. Analytical form of the Schwarz operator and solution to the Riemann-Hilbert problem (1.1) for an arbitrary multiply connected circular domain was obtained through *uniformly* convergent θ_2-Poincaré series in [6] for the continuously differentiable functions $\lambda(t), f(t)$ and in [7] for the Hölder continuous functions. It is worth noting that the principal results were presented by VM in a series of talks reported at the scientific seminar named by F.D. Gakhov (Minsk) in 1991–1992. Due to remarks first of all by S.V. Rogosin and E.I. Zverovich the complete solution to the Riemann-Hilbert problem was "polished". The results were published in [6] and the Russian version was deposited at VINITI, November 3, 1993; Izv. Akad. Nauk Belarus. Ser. Fiz-Mat. 28 pp., Minsk; Deposition No. Z745-V93; RZhMat 1994:3 B117.

The paper by Aleksandrov and Sorokin [1] was devoted to the Schwarz problem for an arbitrary multiply connected circular domain without any separation condition. However, the analytic form of the Schwarz operator was not obtained and advantage of having an exact (closed) form solution was lost. More precisely, the Schwarz problem was reduced via functional equations to an infinite system of linear algebraic equations and to a system of Fredholm integral equations. Truncation of the infinite system was suggested and justified as well. Though the method of truncation can be effective in numeric computations, one can hardly accept that this method yields a closed form solution. Any way it depends on using of the term "closed form solution". A regular infinite system can be considered as an equation with compact operator, i.e., it is no more than a discrete form of a Fredholm integral equation. Therefore, the result [1] can be rather treated as an approximate numerical solution. It is worth noting that the numerical scheme [1] corresponds to frequently used numerical algorithms applied to solution to the Schwarz, Riemann-Hilbert problems and to numerical construction of the Green function, the Schottky-Klein prime function. Such a method works very well when the centers a_k and radii r_k of disks are given numerically [2, 3], but not symbolically. Therefore, the paper [1] contains valuable results but not the exact solution.

It was declared in the papers [15, 16] that the Keldysh-Sedove formulae had been constructed for an arbitrary multiply connected circular domain in terms of the *absolutely and uniformly convergent* series (the name of the series as the Poincaré series was not used. It was called by Golusin's series). It was written in the papers [15, 16] that the proof of *absolute and uniform* convergence was given in [1]. But it is not so. The paper [1] does not contain such a proof and does contain absolute and uniform convergence of another series obtained via a truncation method. Moreover, this assertion from [15, 16] is wrong because of the examples given by Myrberg and Akaza (see [8]) when the absolute convergence of the Poincaré series fails. However, we suppose that formulae from [15, 16] could be correct under the separation condition.

In conclusion, we can say that the mixed (Keldysh-Sedov) problem and the Riemann-Hilbert problem with discontinuous coefficients for an arbitrary multiply connected circular domain has been not completely solved yet. In the present paper, we show how to solve this problem when the coefficient is discontinuous in one of the boundary components. The general case requires long similar investigations and will be presented in separate publications.

References

1. I.A. Aleksandrov, A.S. Sorokin, The problem of Schwarz for multiply connected domains. Sib. Math. Zh. **13**, 971–1001 (1972)
2. R. Balu, T.K. DeLillo, Numerical methods for Riemann-Hilbert problems in multiply connected circle domains. J. Comput. Appl. Math. **307**, 248–261 (2016)
3. T.K. DeLillo, A.R. Elcrat, E.H. Kropf, J.A. Pfaltzgraff, Efficient calculation of Schwarz-Christoffel transformations for multiply connected domains using Laurent series. Comput. Methods Funct. Theory **13**, 307–336 (2013)
4. F.D. Gakhov, *Boundary Value Problems*, 3rd edn. (Nauka, Moscow, 1977, in Russian); Engl. transl. of 1st edn. (Pergamon Press, Oxford, 1966)

5. M.V. Keldysh, L.I. Sedov, Effective solution to some boundary value problems for harmonic functions. Dokl. Akad. Nauk SSSR **16**, 7–10 (1937)
6. V. Mityushev, Solution of the Hilbert boundary value problem for a multiply connected domain. Slupskie Prace Mat-Przyr. **9a**, 33–67 (1994)
7. V. Mityushev, Hilbert boundary value problem for multiply connected domains. Complex Var. **35**, 283–295 (1998)
8. V. Mityushev, Riemann-Hilbert problems for multiply connected domains and circular slit maps. Comput. Methods Funct. Theory **11**, 575–590 (2011)
9. V. Mityushev, Scalar Riemann-Hilbert problem for multiply connected domains, in *Functional Equations in Mathematical Analysis*, ed. by Th.M. Rassias, J. Brzdęk. Springer Optimization and its Applications, vol. 52. (Springer Science+Business Media, LLC, New York, 2012), pp. 599–632. https://doi.org/10.1007/978-1-4614-0055-438
10. V. Mityushev, *$\mathbb{R}$-Linear and Riemann-Hilbert Problems for Multiply Connected Domains*, ed. by S.V. Rogosin, A.A. Koroleva. Advances in Applied Analysis (Birkhäuser, Basel, 2012), pp. 147–176
11. V. Mityushev, *Poincare α-Series for Classical Schottky Groups and its Applications*, ed. by G.V. Milovanović, M.Th. Rassias. Analytic Number Theory, Approximation Theory, and Special Functions (Springer, Berlin, 2014), pp. 827–852
12. V.V. Mityushev, S.V. Rogosin, *Constructive Methods to Linear and Non-linear Boundary Value Problems of the Analytic Function. Theory and Applications*. Monographs and Surveys in Pure and Applied Mathematics (Chapman & Hall/CRC, Boca Raton, 2000)
13. N. Rylko, Fractal local fields in random composites. Appl. Math. Comput. **69**, 247–254 (2015)
14. N. Rylko, Edge effects for heat flux in fibrous composites. Appl. Math. Comput. **70**, 2283–2291 (2015)
15. A.S. Sorokin, The homogeneous Keldysh-Sedov problem for circular multiply connected circular domains in Muskhelishvili's class h_0. Differ. Uravn. **25**, 283–293 (1989)
16. A.S. Sorokin, The Keldysh-Sedov problem for multiply connected circular domains. Sibirsk. Mat. Zh. **36**, 186–202 (1995)

A Boundary Integral Method for the General Conjugation Problem in Multiply Connected Circle Domains

Mohamed M.S. Nasser

Abstract We present a boundary integral method for solving a certain class of Riemann-Hilbert problems known as the general conjugation problem. The method is based on a uniquely solvable boundary integral equation with the generalized Neumann kernel. We present also an alternative proof for the existence and uniqueness of the solution of the general conjugation problem.

Keywords General conjugation problem • Riemann-Hilbert problem • Generalized Neumann kernel

Mathematics Subject Classification (2010) Primary 30E25; Secondary 45B05

1 Introduction

The Riemann-Hilbert problem (RH-problem, for short) is one of the most important classes of boundary value problems for analytic functions. Indeed, the Dirichlet problem, the Neumann problem, the mixed Dirichlet-Neumann problem, the problems of computing the conformal mapping and the external potential flow can be formulated as RH-problems. The RH problem consists of determining all analytic functions in a domain G in the extended complex plane that satisfy a prescribed boundary condition on the boundary $C = \partial G$.

This paper has been presented in: BFA 3rd meeting, Rzeszow, Poland, April 20–23, 2016. The author is grateful to Qatar University for the financial support to attend the meeting and to professor Piotr Drygas, chairman of the organizing committee of the meeting, for the hospitality during the meeting.

M.M.S. Nasser (✉)
Department of Mathematics, Statistics and Physics, Qatar University, P.O. Box 2713, Doha, Qatar
e-mail: mms.nasser@qu.edu.qa

P. Drygaś, S. Rogosin (eds.), *Modern Problems in Applied Analysis*,
Trends in Mathematics, https://doi.org/10.1007/978-3-319-72640-3_11

A boundary integral equation with continuous kernel for solving the RH problem has been derived in [12, 13]. The Kernel of the derived integral equation is a generalization of the well known *Neumann kernel*. So, the new kernel has been called the *generalized Neumann kernel*. The solvability of the boundary integral equation with the generalized Neumann kernel has been studied for simply connected domains with smooth boundaries in [32], for simply connected domains with piecewise smooth boundaries in [22], and for multiply connected domains in [14, 31].

It turns out that the solvability of the boundary integral equation with the generalized Neumann kernel depends on the index κ_j of the coefficient function A of the RH-problem on each boundary component of the boundary of G. However, the solvability of the RH-problem depends on the total index κ of the function A on the whole boundary of G. This raises a difficulty in using the boundary integral equation with the generalized Neumann kernel to solve the RH-problem in multiply connected domains. Such a difficulty does not appear for the simply connected case since the boundary of G consists of only one component. So far, the boundary integral equation with the generalized Neumann kernel has been used to solve the RH-problem in multiply connected domains for special cases of the function A (see [20, Eq. (1.2)]). For such special cases, the boundary integral equation with the generalized Neumann kernel has been used successfully to compute the conformal mapping onto more than 40 canonical domains [15–19, 21, 25] and to solve several boundary value problems such as the Dirichlet problem, the Neumann problem, and the mixed boundary value problem [1, 23, 24].

In this paper, we shall use the boundary integral equation with the generalized Neumann kernel to solve a certain class of RH-problems in multiply connected domains considered by Wegmann [29] and known as the *general conjugation problem*. We shall also use the integral equation to provide an alternative proof for the existence and uniqueness of the solution of the general conjugation problem.

The general conjugation problem considered by Wegmann [29] for multiply connected circular domains is a RH-problem with a particular index. For RH-problems in multiply connected circular domains with general index, it follows from [9, 11] that the existence and uniqueness of a solution for the RH-problem depends on a linear algebraic system which was predicted by Bojarski (see Appendix to Chapter 4 in [27]) and written explicitly in [9, 11]. Further, exact formulas for the solution of the of RH-problem in multiply connected circular domains for general index were given in [9, 11]. These exact formulas were effectively applied in several publications (see e.g., [3, 5, 26]).

2 The Generalized Neumann Kernel

Let G be the unbounded multiply connected circular domain obtained by removing m disks $D_1, \dots, D_m$ from the extended complex plane $\mathbb{C} \cup \{\infty\}$ such that $\infty \in G$ (see Fig. 1). The disk D_j is bounded by the circle $C_j = \partial D_j$ with a center z_j and

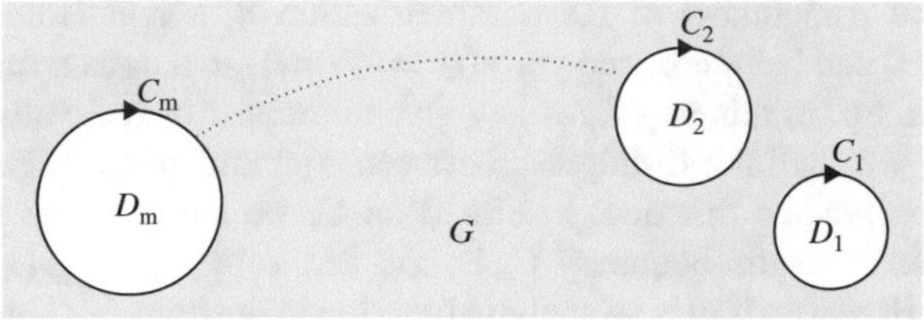

Fig. 1 An unbounded multiply connected circular domain G of connectivity m

radius r_j. We assume that each circle C_j is clockwise oriented and parametrized by

$$\eta_j(t) = z_j + r_j e^{-\mathrm{i}t}, \quad t \in J_j := [0, 2\pi], \quad j = 1, 2, \ldots, m.$$

Let J be the disjoint union of the m intervals $J_1, \ldots, J_m$ which is defined by

$$J = \bigsqcup_{j=1}^{m} J_j = \bigcup_{j=1}^{m} \{(t, j) \ : \ t \in J_j\}. \tag{2.1}$$

The elements of J are order pairs (t, j) where j is an auxiliary index indicating which of the intervals the point t lies in. Thus, the parametrization of the whole boundary $C = \partial D = C_1 \cup C_2 \cup \cdots \cup C_m$ is defined as the complex function η defined on J by

$$\eta(t, j) = \eta_j(t), \quad t \in J_j, \quad j = 1, 2, \ldots, m. \tag{2.2}$$

We assume that for a given t that the auxiliary index j is known, so we replace the pair (t, j) in the left-hand side of (2.2) by t, i.e., for a given point $t \in J$, we always know the interval J_j that contains t. The function η in (2.2) is thus simply written as

$$\eta(t) := \begin{cases} \eta_1(t), & t \in J_1 = [0, 2\pi], \\ \vdots & \\ \eta_m(t), & t \in J_m = [0, 2\pi]. \end{cases} \tag{2.3}$$

Let H denotes the space of all real functions γ in J, whose restriction γ_j to $J_j = [0, 2\pi]$ is a real-valued, 2π-periodic and Hölder continuous function for each $j = 1, \ldots, m$, i.e.,

$$\gamma(t) = \begin{cases} \gamma_1(t), & t \in J_1, \\ \vdots, & \\ \gamma_m(t), & t \in J_m. \end{cases}$$

In view of the smoothness of the parametrization η, a real Hölder continuous function $\hat{\gamma}$ on C can be interpreted via $\gamma(t) := \hat{\gamma}(\eta(t))$, $t \in J$, as a function $\gamma \in H$; and vice versa. So, in this paper, for any given complex or real valued function ϕ defined on C, we shall not distinguish between $\phi(t)$ and $\phi(\eta(t))$. Further, for any complex or real valued function ϕ defined on C, we shall denote the restriction of the function ϕ to the boundary C_j by ϕ_j, i.e., $\phi_j(\eta(t)) = \phi(\eta_j(t))$ for each $j = 1, \ldots, m$. However, if Φ is an analytic function in the domain G, we shall denote the restriction of its values to the boundary C_j by $\Phi_{|j}$, i.e., $\Phi_{|j}(\eta(t)) = \Phi(\eta_j(t))$ for each $j = 1, \ldots, m$.

Let A be a continuously differentiable complex function on C with $A \neq 0$. The generalized Neumann kernel is defined by (see [31] for details)

$$N(s,t) := \frac{1}{\pi} \operatorname{Im} \left(\frac{A(s)}{A(t)} \frac{\dot{\eta}(t)}{\eta(t) - \eta(s)} \right). \tag{2.4}$$

When $A = 1$, the kernel N is the well-known Neumann kernel which appears frequently in the integral equations of potential theory and conformal mapping (see e.g. [8]). We define also the following singular kernel $M(s,t)$ which is closely related to the generalized Neumann kernel $N(s,t)$ [31],

$$M(s,t) := \frac{1}{\pi} \operatorname{Re} \left(\frac{A(s)}{A(t)} \frac{\dot{\eta}(t)}{\eta(t) - \eta(s)} \right). \tag{2.5}$$

Lemma 2.1 ([31])

(a) *The kernel $N(s,t)$ is continuous with*

$$N(t,t) = \frac{1}{\pi} \left(\frac{1}{2} \operatorname{Im} \frac{\ddot{\eta}(t)}{\dot{\eta}(t)} - \operatorname{Im} \frac{\dot{A}(t)}{A(t)} \right). \tag{2.6}$$

(b) *When $s, t \in J_j$ are in the same parameter interval J_j, then*

$$M(s,t) = -\frac{1}{2\pi} \cot \frac{s-t}{2} + M_1(s,t) \tag{2.7}$$

with a continuous kernel M_1 which takes on the diagonal the values

$$M_1(t,t) = \frac{1}{\pi} \left(\frac{1}{2} \operatorname{Re} \frac{\ddot{\eta}(t)}{\dot{\eta}(t)} - \operatorname{Re} \frac{\dot{A}(t)}{A(t)} \right). \tag{2.8}$$

On H we define the Fredholm operator

$$\mathbf{N}\gamma = \int_J N(s,t)\gamma(t)dt$$

and the singular operator

$$\mathbf{M}\gamma = \int_J M(s,t)\gamma(t)dt.$$

Both operators **N** and **M** are bounded on the space H and map H into itself. For more details, see [31]. The identity operator on H is denoted by **I**.

3 The Riemann-Hilbert Problem

The RH-problem for the unbounded multiply connected domain G is defined as follows:
For a given function $\gamma \in H$, search a function Ψ analytic in G and continuous on the closure $\overline{G}$ with $\Psi(\infty) = 0$ such that the boundary values of Ψ satisfy on C the boundary condition

$$\text{Re}[A\Psi] = \gamma. \tag{3.1}$$

The boundary condition in (3.1) is non-homogeneous. When $\gamma \equiv 0$, we have the homogeneous boundary condition

$$\text{Re}[A\Psi] = 0. \tag{3.2}$$

The solvability of the RH problem depends upon the *index* of the function A on the boundary C. The index κ_j of the function A on the circle C_j is the change of the argument of A along the circle C_j divided by 2π. The index κ of the function A on the whole boundary curve C is the sum

$$\kappa = \sum_{j=1}^{m} \kappa_j. \tag{3.3}$$

Remark 3.1 Vekua [27, Eq. (1.2), p. 222], Gakhov [4, Eq. (27.1), p. 208] and Mityushev [9, Eq. (38.3), p. 601] define the RH-problem with $\text{Re}[\overline{A}\Psi] = \gamma$, i.e., with the complex conjugate of the function A. This has the consequence that in some of the later results the index of the function A occurs with the opposite sign as in [4, 9, 27].

We follow [31] and define the space R^+, the spaces of functions γ for which the RH problem (3.1) have solution, by

$$R^+ := \{\gamma \in H : \gamma = \text{Re}[A\Psi] \text{ on } C, \Psi \text{ analytic in } G, \Psi(\infty) = 0\}. \tag{3.4}$$

We define also the space S^+ to be the space of the boundary values of solutions of the homogeneous RH problem, i.e.,

$$S^+ := \{\gamma \in H : \gamma = A\Psi \text{ on } C,\ \Psi \text{ analytic in } G,\ \Psi(\infty) = 0\}. \tag{3.5}$$

To study the solvability of the RH-problem (3.1), we define the following boundary value problem as the homogeneous exterior RH problem on $G^- = D_1 \cup D_2 \cup \cdots \cup D_m$:
Search a function g analytic in G^- and continuous on the closure $\overline{G^-}$ such that the boundary values of g satisfy on C,

$$\mathrm{Re}[Ag] = 0. \tag{3.6}$$

It is clear that G^- is not a domain. In fact, it is the union of m disjoint disks. So, solving the homogeneous exterior RH problem (3.6) is equivalent to solving m RH problems in the disks $D_1, D_2, \ldots, D_m$. The space of the boundary values of solutions of the homogeneous exterior RH problem (3.6) is denoted by S^-, i.e.,

$$S^- := \{\gamma \in H : \gamma = Ag \text{ on } C,\ g \text{ analytic in } G^-\}. \tag{3.7}$$

For $j = 1, 2, \ldots, m$, let S_j^- be the subspace of S^- of real functions $\gamma \in S^-$ such that

$$\gamma(t) = \begin{cases} 0, & t \in J_k,\ k \neq j,\ k = 1, 2, \ldots, m, \\ A_j(t) g(\eta_j(t)), & t \in J_j, \end{cases}$$

where g is analytic in the disk D_j, i.e., g is a solution of the following homogenous RH-problem on the disk D_j,

$$\mathrm{Im}[A_j g] = 0 \quad \text{on} \quad C_j. \tag{3.8}$$

The problem (3.8) is a RH-problem in the bounded simply connected domain D_j and the index of the function A_j on $C_j = \partial D_j$ is κ_j. Then we have from [32, Eq. (29)]

$$\dim(S_j^-) = 2\kappa_j + 1. \tag{3.9}$$

(note that the orientation of the circles C_j is clockwise which changes the sign in [32, Eq. (29)]). Then, we have the following lemmas from [31].

Lemma 3.2 *The space S^- is the direct sum of the subspaces $S_1^-, S_2^-, \ldots, S_m^-$,*

$$S^- = S_1^- \oplus S_2^- \oplus \cdots \oplus S_m^-. \tag{3.10}$$

The space S^- plays a very important rule in using the boundary integral equation with the generalized Neumann kernel to solve the RH-problem (3.1). When the problem is not solvable, the space S^- allows us to find the form of conditions that we should impose on γ to make the problem solvable. For simply connected domains, it was proved in [32, Corollary 3] that the space H has direct sum decomposition

$$H = R^+ \oplus S^-. \tag{3.11}$$

The decomposition (3.11) means that if the RH-problem $\mathrm{Re}[A\Psi] = \gamma$ is not solvable, then there exists a unique function $h \in S^-$ such that the RH-problem $\mathrm{Re}[A\Psi] = \gamma + h$ is solvable. For simply connected domains, the decomposition (3.11) is valid for general index κ of the function A. However, for multiply connected domains, the decomposition (3.11) in general is not correct since we may have $R^+ \cap S^- \neq \{0\}$ (see [31, §10].)

In this paper, we shall consider special cases of the function A for which we can prove that the decomposition (3.11) is valid for multiply connected domains (see Theorem 3.5 below), namely, we assume the index of the function A satisfies

$$\kappa_j \geq 0 \quad \text{for all} \quad j = 1, 2, \ldots, m, \tag{3.12}$$

which implies that $\kappa \geq 0$. RH-problem with such special case of the index has wide applications. For example, our assumption (3.12) on the index are satisfied for the RH-problem used in [15–19, 21, 25] to develop a method for computing the conformal mapping onto more than 40 canonical domains and for the RH-problems studied in [9, 10, 29]. Furthermore, many other boundary value problems such as the Dirichlet problem, the Neumann problem, the mixed boundary value problem, and the Schwarz problem can be reduced to RH-problems whose indexes satisfy our assumption (3.12) (see e.g., [1, 23, 24]).

Under the above assumption (3.12) on the index, we have the following theorem from [14, 31].

Theorem 3.3 *For $\kappa \geq 0$, we have*

$$\mathrm{codim}(R^+) = 2\kappa + m, \quad \dim(S^+) = 0. \tag{3.13}$$

The following lemma follows from (3.9), (3.10) and (3.3).

Lemma 3.4 *Let $\kappa_j \geq 0$ for $j = 1, 2, \ldots, m$, then*

$$\dim(S^-) = 2\kappa + m. \tag{3.14}$$

There is a close connection between RH problems and integral equations with the generalized Neumann kernel. The null-spaces of the operators $\mathbf{I} \pm \mathbf{N}$ are related to the spaces $S^{\pm}$ by (see [31, Theorem 11, Lemma 20])

$$\mathrm{Null}(\mathbf{I} + \mathbf{N}) = S^-, \tag{3.15}$$

$$\mathrm{Null}(\mathbf{I} - \mathbf{N}) = S^+ \oplus W, \tag{3.16}$$

where W is isomorphic via $\mathbf{M}$ to $R^+ \cap S^-$. For the function A defined by (4.3), we have [14, 31]

$$\dim(\mathrm{Null}(\mathbf{I}+\mathbf{N})) = \dim(S^-) = \sum_{j=1}^{m} \max(0, 2\kappa_j + 1) = 2\kappa + m \tag{3.17}$$

and

$$\dim(\mathrm{Null}(\mathbf{I}-\mathbf{N})) = \sum_{j=1}^{m} \max(0, -2\kappa_j - 1) = 0. \tag{3.18}$$

In view of (3.16), Eq. (3.18) implies that $S^+ = W = \{0\}$ (see also (3.13)). Since W isomorphic to $R^+ \cap S^-$, we have also

$$R^+ \cap S^- = \{0\}. \tag{3.19}$$

Theorem 3.5 *Let $\kappa_j \geq 0$ for $j = 1, 2, \ldots, m$, then the space H has the decomposition*

$$H = R^+ \oplus S^- \tag{3.20}$$

Proof Since

$$\mathrm{codim}(R^+) = \dim(S^-) = 2\kappa + m,$$

the direct sum decomposition follows from (3.19). □

It follows from Theorem 3.3 that the non-homogeneous RH problem (3.1) is in general insolvable for $\kappa_j \geq 0$. If it is solvable, then the solution is unique. The following corollary which follows from Theorem 3.5 provides us with a way for modifying the right-hand side of (3.1) to ensure the solvability of the problem.

Corollary 3.6 *Let $\kappa_j \geq 0$ for $j = 1, 2, \ldots, m$, then for any $\gamma \in H$, there exists a unique function $h \in S^-$ such that the following RH problem*

$$\mathrm{Re}[A\Psi] = \gamma + h \tag{3.21}$$

is uniquely solvable.

Solving the RH-problem requires determining both the analytic function Ψ as well as the real function h. This can be done easily using the boundary integral equations with the generalized Neumann kernel as in the following theorem.

Theorem 3.7 *Let $\kappa_j \geq 0$ for $j = 1, 2, \ldots, m$. For any given $\gamma \in H$, let μ be the unique solution of the integral equation*

$$\mu - \boldsymbol{N}\mu = -\boldsymbol{M}\gamma. \tag{3.22}$$

Then the boundary values of the unique solution of the RH problem (3.21) is given by

$$A\Psi = \gamma + h + \mathrm{i}\mu \tag{3.23}$$

and the function h is given by

$$h = [\boldsymbol{M}\mu - (\boldsymbol{I} - \boldsymbol{N})\gamma]/2. \tag{3.24}$$

Proof The theorem can be proved using the same argument as in the proof of [15, Theorem 2]. □

The uniqueness of the solution of the integral equation (3.22) follows from the Fredholm alternative theorem since $\dim(\mathrm{Null}(\mathbf{I} - \mathbf{N})) = 0$. The advantages of Theorem 3.7 are that it, based on the integral equation (3.22), provides us with formulas for computing the real function h necessary for the solvability of the RH problem as well as the solution Ψ of the RH problem.

4 The General Conjugation Problem

In this section, we shall consider a very important certain class of RH problems which has been considered by Wegmann [29]. The indexes of this class of RH problems satisfied our assumption (3.12). This class has been considered by many researchers and has many applications [2, 11, 28–30].

Wegmann [29] proves the following theorem.

Theorem 4.1 *For any integer $\ell \geq 0$ and for any sufficiently smooth function γ on the boundary of G, the RH problem*

$$\mathrm{Re}\left[e^{\mathrm{i}\lambda_j} e^{\mathrm{i}\ell t}\Psi_{|j}(\eta(t)) + (a_{j\ell} + \mathrm{i}b_{j\ell})e^{\mathrm{i}\ell t} + \cdots + (a_{j1} + \mathrm{i}b_{j1})e^{\mathrm{i}t} + a_{j0}\right] = \gamma_j(t), \tag{4.1}$$

has a unique solution consisting of an analytic function Ψ in G with $\Psi(\infty) = 0$ and $(2\ell + 1)m$ real numbers $a_{j0}, a_{j1}, \ldots, a_{j\ell}, b_{j1}, \ldots, b_{j\ell}$ for $j = 1, \ldots, m$.

Wegmann [29] called the RH problem (4.1) as the *general conjugation problem* since the case $\ell = 0$ describes the problem of finding the conjugate harmonic function of a harmonic function with boundary values γ_j [29]. The general conjugation problem (4.1) has been solved by Wegmann [29] using the method of *successive conjugation* which reduces the problem (4.1) to a sequence of RH problems on the circles C_j. This method has been first applied by Halsey [7] for $\ell = 0$. Applications of this problem to compute conformal mapping have been given in [28, 30] for $\ell = 1$ and in [2, 30] for $\ell = 0$.

In this paper, we shall present a method for solving the general conjugation problem (4.1) for any $\ell \geq 0$. The method is based on the boundary integral equation with the generalized Neumann kernel (3.22). However, we shall first rewrite the problem in a form suitable for using the integral equation.

The boundary condition (4.1) can be written as

$$\mathrm{Re}\left[e^{\mathrm{i}\lambda_k}e^{\mathrm{i}\ell t}\Psi_{|j}\right] = \gamma_j - a_{j0} - \sum_{k=1}^{\ell}\mathrm{Re}\left[(a_{jk}+\mathrm{i}b_{jk})e^{\mathrm{i}kt}\right], \quad j=1,2,\ldots,m. \tag{4.2}$$

Let A be defined by

$$A(t) := \begin{cases} e^{\mathrm{i}\lambda_1}e^{\mathrm{i}\ell t}, & t\in J_1=[0,2\pi], \\ \vdots & \\ e^{\mathrm{i}\lambda_m}e^{\mathrm{i}\ell t}, & t\in J_m=[0,2\pi]. \end{cases} \tag{4.3}$$

Let also ψ be a function defined on the boundary C where its values on C_j is given by

$$\psi_j(t) = -a_{j0} - \sum_{k=1}^{\ell}\mathrm{Re}\left[(a_{jk}+\mathrm{i}b_{jk})e^{\mathrm{i}kt}\right], \quad j=1,2,\ldots,m. \tag{4.4}$$

Hence, the general conjugation problem (4.1) can be written as the RH-problem

$$\mathrm{Re}\,[A\Psi] = \gamma + \psi. \tag{4.5}$$

By the existence and uniqueness of the solution of the general conjugation problem (4.1), the RH-problem (4.5) has a unique solution. Remember that solving the RH-problem (4.5) requires determining the analytic function Ψ and the unknown real function ψ. Determining the function ψ is equivalent to determining the $m(2\ell+1)$ real constants a_{j0}, a_{jk}, and b_{jk} in (4.1) for $j=1,\ldots,m$, $k=1,\ldots,\ell$.

The index of the function A given by (4.3) is

$$\kappa_j = \ell \geq 0, \quad j=1,\ldots,m,$$

and hence the total index is $\kappa = m\ell \geq 0$. Thus our assumption (3.12) is satisfied. We shall use the integral equation with the generalized Neumann kernel (3.22) to determine the analytic function Ψ as well as the real function ψ in (4.5). However, we need to show first that the function ψ is indeed in S^-. This can be proved by finding the explicit form of the functions of the space S^- which will be given in the next section.

5 The Space S^-

To find the explicit form of the space S^-, we need the following theorem from [4, § 29.3].

Theorem 5.1 *Let $D = \{z : |z| < 1\}$ and the boundary $C = \partial D$ is the unit circle parametrized by $\zeta(t) = e^{-\mathrm{i}t}$, $t \in [0, 2\pi]$. For $\ell \geq 0$, the solution of the following homogenous RH problem on D,*

$$\mathrm{Re}\left[\zeta(t)^{-\ell} g(\zeta(t))\right] = 0, \quad \zeta(t) \in C,$$

is given for $z \in \overline{D}$ by

$$g(z) = z^{\ell}\left[\mathrm{i}c_0 + \frac{1}{2}\sum_{k=1}^{\ell}\left(c_k z^k - \overline{c_k} z^{-k}\right)\right]$$

where c_0 is an arbitrary real constant and $c_1, \ldots, c_\ell$ are arbitrary complex constants.

Lemma 5.2 *Let $h \in H$. Then $h \in S_j^-$ if and only if it has the form*

$$h(t) = \begin{cases} 0, & \text{on } C_k \text{ for } k \neq j, \\ \beta_{j0} + \sum_{k=1}^{\ell} \mathrm{Re}\left[(\beta_{jk} + \mathrm{i}\alpha_{jk})e^{\mathrm{i}kt}\right], & \text{on } C_j, \end{cases} \tag{5.1}$$

where $\beta_{j0}, \beta_{jk}, \alpha_{jk}, j = 1, 2, \ldots, m, k = 1, 2, \ldots, \ell$, are real constants.

Proof By the definition of the space S_j^-, a function $h \in S_j^-$ if and only if

$$h(t) = \begin{cases} 0, & \text{on } C_k \text{ for } k \neq j, \\ A_j(t) g(\eta_j(t)), & \text{on } C_j, \end{cases} \tag{5.2}$$

where g is analytic in D_j and $A_j(t)$ is the restriction of the function A given by (4.3) to J_j. Using the definition (4.3) of the function A, the function $\mathrm{i}g$ is a solution of the homogeneous RH problem

$$\mathrm{Re}[e^{\mathrm{i}\lambda_j} e^{\mathrm{i}\ell t}(\mathrm{i}g(\eta_j(t)))] = 0 \tag{5.3}$$

on the disk D_j. We define an analytic function F in D_j by

$$F(z) = \mathrm{i}e^{\mathrm{i}\lambda_j} g(z).$$

Then F is a solution of the homogeneous RH problem

$$\mathrm{Re}[e^{\mathrm{i}\ell t} F(\eta_j(t))] = 0 \tag{5.4}$$

on the disk D_j. The function

$$\omega(z) = \frac{z - z_j}{r_j}$$

is analytic on D_j and maps the circle C_j onto the unit circle and

$$\omega(\eta_j(t)) = \frac{\eta_j(t) - z_j}{r_j} = e^{-\mathrm{i}t} = \zeta(t)$$

is the parametrization of the unit circle. Let the function $\hat{F}$ be defined on the unit disk D by

$$\hat{F}(z) = F(\omega^{-1}(z)) = F(z_j + r_j z).$$

Then, it follows from (5.4) that $\hat{F}$ is a solution of the homogeneous RH problem

$$\mathrm{Re}\left[\zeta(t)^{-\ell}\hat{F}(\zeta(t))\right] = 0 \tag{5.5}$$

in the unit disk D. Hence Theorem 5.1 implies that

$$\zeta(t)^{-\ell}\hat{F}(\zeta(t)) = \mathrm{i}\beta_{j0} + \frac{1}{2}\sum_{k=1}^{\ell}[(\alpha_{jk} + \mathrm{i}\beta_{jk})\zeta(t)^k - (\alpha_{jk} - \mathrm{i}\beta_{jk})\zeta(t)^{-k}]$$

where $\beta_{j0}, \beta_{jk}, \alpha_{jk}$, $j = 1, 2, \ldots, m$, $k = 1, 2, \ldots, \ell$, are arbitrary real constants. Since $\zeta(t) = e^{-\mathrm{i}t}$ and $\hat{F}(\zeta(t)) = F(\eta_j(t)) = \mathrm{i}e^{\mathrm{i}\lambda_j}g(\eta_j(t))$, we obtain

$$\mathrm{i}e^{\mathrm{i}\ell t}e^{\mathrm{i}\lambda_j}g(\eta_j(t)) = \mathrm{i}\beta_{j0} + \frac{1}{2}\sum_{k=1}^{\ell}[(\alpha_{jk} + \mathrm{i}\beta_{jk})e^{-\mathrm{i}kt} - (\alpha_{jk} - \mathrm{i}\beta_{jk})e^{\mathrm{i}kt}]. \tag{5.6}$$

Since $h_j(t) = e^{\mathrm{i}\ell t}e^{\mathrm{i}\lambda_j}g(\eta_j(t))$, (5.6) implies that

$$h_j(t) = \beta_{j0} + \frac{1}{2}\sum_{k=1}^{\ell}\left[(\beta_{jk} - \mathrm{i}\alpha_{jk})e^{-\mathrm{i}kt} + (\beta_{jk} + \mathrm{i}\alpha_{jk})e^{\mathrm{i}kt}\right]$$

which can be written as

$$h_j(t) = \beta_{j0} + \sum_{k=1}^{\ell}\mathrm{Re}\left[(\beta_{jk} + \mathrm{i}\alpha_{jk})e^{\mathrm{i}kt}\right]. \tag{5.7}$$

Hence (5.1) follows from (5.2) and (5.7). □

Theorem 5.3 *Let $h \in H$. Then $h \in S^-$ if and only if it has the form*

$$h_j(t) = \beta_{j0} + \sum_{k=1}^{\ell}\mathrm{Re}\left[(\beta_{jk} + \mathrm{i}\alpha_{jk})e^{\mathrm{i}kt}\right], \tag{5.8}$$

where $\beta_{j0}, \beta_{jk}, \alpha_{jk}$, $j = 1, 2, \ldots, m$, $k = 1, 2, \ldots, \ell$, *are* $(1 + 2\ell)m$ *real constants.*

Proof The proof follows from Lemmas 3.2 and 5.2. □

In view of (5.8), it is clear that the function ψ in (4.4) is in the space S^-. By the uniqueness of the function ψ in (4.5) and the function h in (3.21), we conclude that the functions ψ and h are identical, i.e., $\psi \equiv h$ where h is given by (3.24).

Corollary 5.4 *Let* $h \in H$. *If* $\ell = 0$, *then* $h \in S^-$ *if and only if it has the form*

$$h_j(t) = \beta_{j0}, \tag{5.9}$$

where β_{j0}, $j = 1, 2, \ldots, m$, $k = 1, 2, \ldots, \ell$, *are* m *real constants.*

The above corollary is correct for general multiply connected domains whose boundaries are smooth Jordan curves (see [23]). However, for $\ell > 0$, the Formula (5.8) is valid only for multiply connected domains with circular boundaries. In view of the existence of a conformal mapping $w = \Upsilon(z)$ from general multiply connected domains to multiply connected circular domains [6], Theorem 5.3 can be extended to general multiply connected domains. For the latter case, the Formula (5.8) will involve the conformal mapping Υ (see e.g., [4, Eq. (29.27), p. 226] for simply connected domains case).

6 Solving the General Conjugation Problem

Since the function ψ in (4.5) and the function h given by (3.24) are identical, the RH-problem (4.2) or equivalently the general conjugation problem (4.1) can be solved by the integral equation with the generalized Neumann kernel (3.22) as in the following theorem.

Theorem 6.1 *For any* $\gamma \in H$, *the boundary values of the unique solution* Ψ *of the general conjugation problem* (4.1) *are given by*

$$\Psi(\eta(t)) = \frac{\gamma(t) + h(t) + \mathrm{i}\mu(t)}{A(t)} \tag{6.1}$$

where $A(t)$ *is defined by* (4.3), $\mu(t)$ *is the unique solution of the integral equation* (3.22) *and the function* $h(t)$ *is given by* (3.24).

To evaluate the $(2\ell + 1)m$ unknown real constants a_{j0}, a_{jk} and b_{jk} in (4.1), $j = 1, \ldots, m$, $k = 1, \ldots, \ell$, we rewrite the function $\psi_j(t)$ in (4.4) as

$$\psi_j(t) = -\sum_{k=0}^{\ell} a_{jk} \cos kt + \sum_{k=1}^{\ell} b_{jk} \sin kt, \quad j = 1, 2, \ldots, m. \tag{6.2}$$

By obtaining the real function h from (3.24) and since $\psi = h$, we have the following theorem.

Theorem 6.2 *The values of the* $(2\ell+1)m$ *unknown real constants* a_{jk} *and* b_{jk}*in* (4.1) *are given by*

$$a_{j0} = -\frac{1}{2\pi}\int_0^{2\pi} h_j(t)dt,$$

$$a_{jk} = -\frac{1}{\pi}\int_0^{2\pi} h_j(t)\cos ktdt,$$

$$b_{jk} = \frac{1}{\pi}\int_0^{2\pi} h_j(t)\sin ktdt,$$

for $j = 1,\dots,m$ *and* $k = 1,\dots,\ell$ *where the function* $h(t)$ *is given by* (3.24).

In the above two theorems, we have used the integral equation (3.22) to develop a method for solving the RH problem (4.1). We can also use Theorem 5.3 and Corollary 3.6 to provide alternative proof for Theorem 4.1 for any $\gamma \in H$.

Alternative proof of Theorem 4.1 For any integer $\ell \geq 0$, let the function A be given by (4.3). Then for any functions $\gamma \in H$, it follows from Corollary 3.6 that a unique function $h \in S^-$ exists such that the RH problem

$$\mathrm{Re}[Af] = \gamma + h \tag{6.3}$$

is uniquely solvable. The function h is given by (3.24) where μ is the unique solution of the integral equation (3.22). Then Theorem 5.3 implies that unique values of the $(2\ell + 1)m$ real constants $\beta_{j0}, \beta_{jk}, \alpha_{jk}$, $j = 1, 2, \dots, m$, $k = 1, 2, \dots, \ell$, exist such that

$$h_j(t) = \beta_{j0} + \sum_{k=1}^{\ell} \mathrm{Re}\left[(\beta_{jk} + \mathrm{i}\alpha_{jk})e^{ikt}\right].$$

Hence the RH problem (6.3) can be written as

$$\mathrm{Re}\left(A_j f_{|j} - \beta_{j0} - \sum_{k=1}^{\ell} \mathrm{Re}\left[(\beta_{jk} + \mathrm{i}\alpha_{jk})e^{ikt}\right]\right) = \gamma_j, \tag{6.4}$$

which in view of the definition (4.3) of the function A gives the proof of Theorem 4.1 where $\Psi = f$ and $a_{j0} = -\beta_{j0}, a_{jk} = -\beta_{jk}, b_{jk} = -\alpha_{jk}$, $j = 1, 2, \dots, m$, $k = 1, 2, \dots, \ell$. □

7 Concluding Remarks

We have presented a boundary integral method for solving the general conjugation problem which is a certain class of Riemann-Hilbert problems. The method is based on a uniquely solvable boundary integral equation with the generalized Neumann kernel. We have also presented an alternative proof of Theorem 4.1 which has been proved previously in [29]. The numerical implementation of the above proposed method will be presented in future works.

Acknowledgements The author would like to thank the anonymous referee for his valuable comments and suggestions which improved the presentation of this paper. The financial support from Qatar University Internal Grant QUUG-CAS-DMSP-15\16-27 is gratefully acknowledged.

References

1. S.A.A. Al-Hatemi, A.H.M. Murid, M.M.S. Nasser, A boundary integral equation with the generalized Neumann kernel for a mixed boundary value problem in unbounded multiply connected regions. Bound. Value Probl. **2013**, 54 (2013)
2. R. Balu, T.K. DeLillo, Numerical methods for Riemann-Hilbert problems in multiply connected circle domains. J. Comput. Appl. Math. **307**, 248–261 (2016)
3. R. Czapla, W. Nawalaniec, V. Mityushev, Effective conductivity of random two-dimensional composites with circular non-overlapping inclusions. Comput. Mater. Sci. **63**, 118–126 (2012)
4. F.D. Gakhov, *Boundary Value Problems* (Pergamon Press, Oxford, 1966)
5. S. Gluzman, V. Mityushev, W. Nawalaniec, G. Sokal, Random composite: stirred or shaken? Arch. Mech. **68**(3), 229–241 (2016)
6. G.M. Goluzin, *Geometric Theory of Functions of a Complex Variable* (American Mathematical Society, Providence, RI, 1969)
7. N.D. Halsey, Potential flow analysis of multielement airfoils using conformal mapping. AIAA J. **17**, 1281–1288 (1979)
8. P. Henrici, *Applied and Computational Complex Analysis*, vol. 3 (Wiley, New York, 1986)
9. V.V. Mityushev, Scalar Riemann-Hilbert problem for multiply connected domains, in *Functional Equations in Mathematical Analysis*, ed. by Th.M. Rassias, J. Brzdek (Springer, Berlin, 2011), pp. 599–632
10. V. Mityushev, $\mathbb{R}$-linear and Riemann-Hilbert problems for multiply connected domains, in *Advances in Applied Analysis*, ed. by S.V. Rogosin, A.A. Koroleva (Birkhäuser, Basel, 2012), pp. 147–176
11. V. Mityushev, S. Rogosin, *Constructive Methods for Linear and Nonlinear Boundary Value Problems for Analytic Functions* (Chapman & Hall, London, 2000)
12. A.H.M. Murid, M.M.S. Nasser, Eigenproblem of the generalized Neumann kernel. Bull. Malays. Math. Sci. Soc. **26**(2), 13–33 (2003)
13. A.H.M. Murid, M.R.M. Razali, M.M.S. Nasser, Solving Riemann problem using Fredholm integral equation of the second kind, in *Proceeding of Simposium Kebangsaan Sains Matematik Ke-10*, ed. by A.H.M. Murid (UTM, Johor, 2002), pp. 171–178
14. M.M.S. Nasser, The Riemann-Hilbert problem and the generalized Neumann kernel on unbounded multiply connected regions. Univ. Res. (IBB Univ. J.) **20**, 47–60 (2009)
15. M.M.S. Nasser, A boundary integral equation for conformal mapping of bounded multiply connected regions. Comput. Methods Funct. Theory **9**, 127–143 (2009)
16. M.M.S. Nasser, Numerical conformal mapping via a boundary integral equation with the generalized Neumann kernel. SIAM J. Sci. Comput. **31**(3), 1695–1715 (2009)
17. M.M.S. Nasser, Numerical conformal mapping of multiply connected regions onto the second, third and fourth categories of Koebe's canonical slit domains. J. Math. Anal. Appl. **382**, 47–56 (2011)
18. M.M.S. Nasser, Numerical conformal mapping of multiply connected regions onto the fifth category of Koebe's canonical slit regions. J. Math. Anal. Appl. **398**, 729–743 (2013)

19. M.M.S. Nasser, Fast computation of the circular map. Comput. Methods Funct. Theory **15**(2), 187–223 (2015)
20. M.M.S. Nasser, Fast solution of boundary integral equations with the generalized Neumann kernel. Electron. Trans. Numer. Anal. **44**, 189–229 (2015)
21. M.M.S. Nasser, F.A.A. Al-Shihri, A fast boundary integral equation method for conformal mapping of multiply connected regions. SIAM J. Sci. Comput. **35**(3), A1736–A1760 (2013)
22. M.M.S. Nasser, A.H.M. Murid, Z. Zamzamir, A boundary integral method for the Riemann–Hilbert problem in domains with corners. Complex Var. Elliptic Equ. **53**, 989–1008 (2008)
23. M.M.S. Nasser, A.H.M. Murid, M. Ismail, E.M.A. Alejaily, Boundary integral equations with the generalized Neumann kernel for Laplace's equation in multiply connected regions. Appl. Math. Comput. **217**, 4710–4727 (2011)
24. M.M.S. Nasser, A.H.M. Murid, S.A.A. Al-Hatemi, A boundary integral equation with the generalized Neumann kernel for a certain class of mixed boundary value problem. J. Appl. Math. **2012**, 254123, 17 pp (2012)
25. M.M.S. Nasser, J. Liesen, O. Sète, Numerical computation of the conformal map onto lemniscatic domains. Comput. Methods Funct. Theory **16**(4), 609–635 (2016)
26. S.D. Ryan, V. Mityushev, V. Vinokur, L. Berlyand, Rayleigh approximation to ground state of the Bose and Coulomb glasses. Sci. Rep. Nat. Publ. **5**, 7821 (2015)
27. I.N. Vekua, *Generalized Analytic Functions* (Pergamon Press, London, 1992)
28. R. Wegmann, Fast conformal mapping of multiply connected regions. J. Comput. Appl. Math. **130**, 119–138 (2001)
29. R. Wegmann, Constructive solution of a certain class of Riemann–Hilbert problems on multiply connected circular regions. J. Comput. Appl. Math. **130**, 139–161 (2001)
30. R. Wegmann, Methods for numerical conformal mapping, in *Handbook of Complex Analysis, Geometric Function Theory*, vol. 2, ed. by R. Kuehnau (Elsevier, Amsterdam, 2005), pp. 351–477
31. R. Wegmann, M.M.S. Nasser, The Riemann-Hilbert problem and the generalized Neumann kernel on multiply connected regions. J. Comput. Appl. Math. **214**, 36–57 (2008)
32. R. Wegmann, A.H.M. Murid, M.M.S. Nasser, The Riemann-Hilbert problem and the generalized Neumann kernel. J. Comput. Appl. Math. **182**, 388–415 (2005)

Pseudo-Differential Operators on Manifolds with a Singular Boundary

Vladimir B. Vasilyev

Abstract The aim of this work is to describe new interesting examples of non-smooth manifolds and elliptic pseudo-differential operators acting in functional spaces on such manifolds. Fredholm properties for these operators are studied by factorization methods, and these are based on several complex variables.

Keywords Pseudo-differential operator • Local representative • Bochner operator • Wave factorization

Mathematics Subject Classification (2010) Primary 47G30; Secondary 32A07

1 Introduction

In 1990s the author has started to develop a new approach to pseudo-differential equations and boundary value problems on smooth manifolds with a non-smooth boundary. This approach takes its origin from Vishik–Eskin's theory of boundary value problems for manifolds with a smooth boundary [3]. The author has extended a factorization idea to non-smooth situation to apply it for describing Fredholm conditions for elliptic pseudo-differential equations and boundary value problems on manifolds with a non-smooth boundary. This studying was completed in general for a two-dimensional case [11] but last years the author has found new interesting constructions for a multidimensional case also [12–15]. A special case is the author's paper [10] in which it was done for Calderon–Zygmund operators only.

A basic problem is to describe Fredholm conditions for the equation

$$(Au)(x) = v(x), \quad x \in M,$$

V.B. Vasilyev (✉)
Chair of Differential Equations, Belgorod National Research University, Studencheskaya 14/1, Belgorod 308007, Russia
e-mail: vbv57@inbox.ru

P. Drygaś, S. Rogosin (eds.), *Modern Problems in Applied Analysis*,
Trends in Mathematics, https://doi.org/10.1007/978-3-319-72640-3_12

where M is a compact manifold with a boundary (non-smooth as a rule), A is a pseudo-differential operator acting in certain functional spaces on M. We use Sobolev–Slobodetskii spaces $H^s(M)$, $s \in \mathbb{R}$, for studying these properties [11]. Such spaces are introduced locally using a partition of identity [3].

Some words on other approaches to the problem and related topics [2, 4–8]. Richard Melrose's programme was declared in ICM-90 [5], and it is devoted to algebraical and topological aspects of the problem, this way is reserved by papers [6, 7]. Other kind of papers was written by analysts. So, B.-W. Schulze and his group (see for example [2]) deals with general pseudo-differential operators on manifolds with cones and wedges.

"In all papers the conical domain (see Fig. 1 below) is treated as the direct product of a circle and a half-axis, then they apply the Mellin transform on half-axis, and the initial problem is reduced to a problem in a domain with a smooth boundary with operator-valued symbol. That follows further it is like the generalization of well-known results on operator symbol case. Of course, the my approach is generalization also, but it is a generalization on dimension space, and the principal difference is that I don't divide the cone, and it's treated as an emergent thing" [14]. The last is related also to papers V. Mazya, B. Plamenevskii and others mentioned in [14].

Another kind of "analytical" papers deals with very "simple" operators and boundary value problems on manifolds with very "bad" boundary (see for example [4]). The main aim for them is developing the classical Fredholm theory as far as it is possible using potential theory.

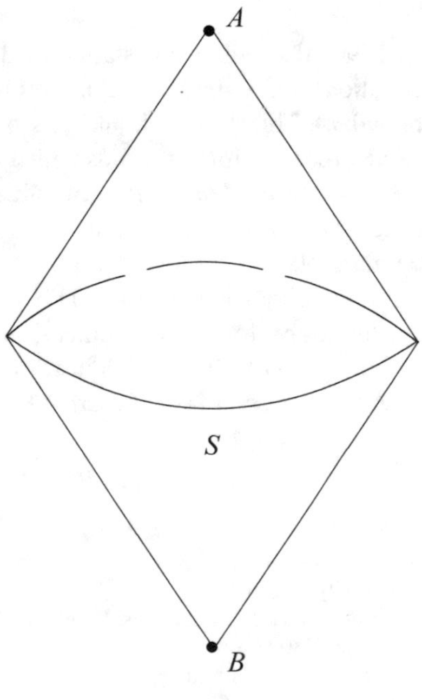

Fig. 1 Simplest manifold with a non-smooth boundary

Finally recent paper [8] contains a certain new generalizations for statements of boundary value problems without smoothness requirement on boundary of a domain.

> ...we develop the global symbolic calculus of pseudo-differential operators generated by a boundary value problem for a given (not necessarily self-adjoint or elliptic) differential operator. For this, we also establish elements of a non-self-adjoint distribution theory and the corresponding biorthogonal Fourier analysis. There are no assumptions on the regularity of the boundary which is allowed to have arbitrary singularities. We give applications of the developed analysis to obtain a priori estimates for solutions of boundary value problems that are elliptic within the constructed calculus. [8]

2 Non-Smooth Manifolds and Local Representatives

Let it will not be a strange thing but in this section we'll define the declared manifold by operators which live on this manifold.

2.1 *Pseudo-differential Operators*

The main object of the paper is a linear bounded operator $A : H^{s_1}(M) \to H^{s_2}(M)$ which is called a pseudo-differential operator under following assumptions. We'll suppose that operator A is composed by a certain operator-function $A(x_0), x_0 \in \overline{M}$, so that for arbitrary smooth functions φ, ψ on M with supports concentrated in small neighborhoods U, V $(U \subset V)$ of x the following representation for the operator A

$$\varphi \cdot A \cdot \psi = \hat{\varphi} \cdot (A_{x_0} + T_{x_0})\hat{\psi},$$

holds, where A_{x_0} is an operator defined by formula

$$(A_{x_0}u)(x) = \int_D \int_{\mathbb{R}^m} e^{i\xi\cdot(x-y)}\tilde{A}(\eta^{-1}(x_0), \xi)\tilde{u}(\xi)d\xi dy, \quad x \in D,$$

$\eta : V \to D \subset \mathbb{R}^m$ is a local diffeomorphism, $\tilde{A}(x, \xi)$ is a certain function defined on $V \times \mathbb{R}^m$, $\hat{\varphi} = \varphi \circ \eta^{-1}, \hat{\psi} = \psi \circ \eta^{-1}$, $T_{x_0} : H^{s_1}(\mathbb{R}^m) \to H^{s_2}(\mathbb{R}^m)$ is a compact operator, $\tilde{u}(\xi)$ denotes the Fourier transform in m-dimensional space

$$\tilde{u}(\xi) = \int_{\mathbb{R}^m} e^{ix\cdot\xi}u(x)dx,$$

$\forall u \in S(\mathbb{R}^m)$ (Schwartz class of infinitely differentiable rapidly decreasing functions at infinity).

Remark 2.1 Generally speaking the domain D depends on the point x_0 but we'll see below in a lot of cases it isn't essential.

Definition 2.2 The operator $A_{x_0} : H^{s_1}(D) \to H^{s_2}(D)$ is called a **local representative** of the operator $A : H^{s_1}(M) \to H^{s_2}(M)$ at the point $x_0 \in \overline{M}$, and domain D is called a **canonical** domain.

A structure of the set D can be different in dependence on a placement of the point x_0. For inner points $\overset{\circ}{M}$ we have $D = \mathbb{R}^m$, for points of smoothness on ∂M we have $D = \mathbb{R}^m_+$, for conical points $D = C^a_+$ and so on. One has painted a simple example of a non-smooth manifold M on the Fig. 1. There are distinct types of local representatives in dependence on a point kind. These local representatives are defined by different formulas for inner points $\overset{\circ}{M}$, for points of smoothness on ∂M, for points on S which is a smooth edge of a wedge, and separately for conical points A and B. Below we'll describe these local representatives and methods for their studying.

Definition 2.3 We say that a manifold M has a non-smooth boundary if there is at least one local representative of a pseudo-differential operator for which $D \neq \mathbb{R}^m, \mathbb{R}^m_+$, and the function $\sigma(x_0, \xi) = \tilde{A}(\eta^{-1}(x_0), \xi)$ defined on $\overline{M} \times \mathbb{R}^m$ is called a **local symbol** of a pseudo-differential operator A at the point x_0.

2.2 *Canonical Domains: A Half-Space, a Cone, a Wedge*

Since the definition of a pseudo-differential operator is a local and we use the "freezing coefficients principle" or, in other words, "local principle" then we'll omit a pole x_0 in a symbol of a pseudo-differential operator. Thus we have following types of local operators related to an initial pseudo-differential operator A. First it is the operator

$$u(x) \longmapsto \int\limits_{\mathbb{R}^m}\int\limits_{\mathbb{R}^m} \tilde{A}(\cdot, \xi)u(y)e^{i(x-y)\cdot\xi}d\xi dy, \quad x \in \mathbb{R}^m. \tag{2.1}$$

for a point $x_0 \in \overset{\circ}{M}$.

If $x_0 \in \partial M$ and x_0 is a smoothness point then we need another formula

$$u(x) \longmapsto \int\limits_{\mathbb{R}^m_+}\int\limits_{\mathbb{R}^m} \tilde{A}(\cdot, \xi)u(y)e^{i(x-y)\cdot\xi}d\xi dy, \quad x \in \mathbb{R}^m_+. \tag{2.2}$$

For invertibility of such an operator with symbol $\tilde{A}(\cdot, \xi)$ not depending on a spatial variable x_0 one can apply the theory of the classical Riemann boundary value problem for upper and lower complex half-planes with a parameter $\xi' = (\xi_1, \ldots, \xi_{m-1})$. This step was systematically studied in the book [3]. But if the boundary ∂M has at least one conical point, this approach is not effective.

A conical point x_0 at the boundary is such a point for which its neighborhood is diffeomorphic to the cone $C^a_+ = \{x \in \mathbb{R}^m : x_m > a|x'|, x' = (x_1, \ldots, x_{m-1}), a > 0\}$, hence the local definition for pseudo-differential operator near the conical point is the following

$$u(x) \longmapsto \int\limits_{C^a_+} \int\limits_{\mathbb{R}^m} \tilde{A}(\cdot, \xi) u(y) e^{i(x-y)\cdot\xi} d\xi dy, \quad x \in C^a_+. \tag{2.3}$$

A k-wedge point x_0 at the boundary is such a point for which its neighborhood is diffeomorphic to the wedge $W^{a_k,k}_+ = \{x \in \mathbb{R}^m : x = (x'', x', x_m), x'' = (x_1, \cdots, x_k), x' = (x_{k+1}, \cdots, x_{m-1}), x_m > a_k|x'|, a_k > 0\}$. In other words $W^{a_k,k}_+ = \mathbb{R}^k \times C^{a_k}_+$, where $C^{a_k}_+$ is a cone in $\mathbb{R}^{m-k}$. Hence the local definition for pseudo-differential operator near the k-wedge point is the following

$$u(x) \longmapsto \int\limits_{W^{a_k,k}_+} \int\limits_{\mathbb{R}^m} \tilde{A}(\cdot, \xi) u(y) e^{i(x-y)\cdot\xi} d\xi dy, \quad x \in W^a_+. \tag{2.4}$$

To study an invertibility property for the operator (2.3), (2.4) the author has introduced the concept of wave factorization for an elliptic symbol near a singular boundary point [10, 11] and using this property has described Fredholm properties for an equation with the operator (2.3), (2.4).

2.3 Class of Symbols

To describe invertibility conditions for operators (2.1)–(2.3) we need to fix a class of local symbols under consideration.

Definition 2.4 A local symbol belongs to the class S_α if it satisfies the following condition

$$|\sigma(x_0, \xi)| \sim (1 + |\xi|)^\alpha, \quad \forall x_0 \in \overline{M}, \quad \xi \in \mathbb{R}^m.$$

The number α is called an order of a pseudo-differential operator.

Such symbols and corresponding operators we call **elliptic** ones.

According to [3] such operators with local symbols from S_α are linear bounded operators acting from $H^s(D)$ to $H^s(D)$, and everywhere below we consider only symbols from the class S_α.

2.4 Main Theorem

Theorem 2.5 *An elliptic pseudo-differential operator $A : H^s(M) \to H^{s-\alpha}(M)$ with continuous local symbol has a Fredholm property iff all local representatives A_{x_0} : $H^s(D) \to H^{s-\alpha}(D)$ are invertible.*

Proof Since the definition of a pseudo-differential operator given above assume that an operator is defined locally then a local principle [9] implies this assertion. □

2.5 Wave Factorization and Invertibility of Local Operators

We'll give the definition of a wave factorization with respect to a k-dimensional wedge because a cone is a particular case of a wedge $W_+^{a_0,0} = C_+^{a_0}$.

Let $\overset{*}{C_+^a} = \{x \in \mathbb{R}^m : ax_m > |x'|\}$ be a conjugate cone, and $T(\overset{*}{C_+^a})$ be a radial tube domain over the cone $\overset{*}{C_+^a}$ [1, 11, 16], it is a subset of $\mathbb{C}^m$ of the following type $T(\overset{*}{C_+^a}) = \mathbb{R}^m + i\overset{*}{C_+^a}$.

Definition 2.6 Wave factorization of a local elliptic symbol $\sigma(x_0, \xi)$ with respect to the wedge $W_+^{a_k,k}$ is called its representation in the form

$$\sigma(x_0, \xi) = \sigma_{\neq}(x_0, \xi) \cdot \sigma_{=}(x_0, \xi),$$

where factors admit analytic continuation into radial tube domains $T(\pm \overset{*}{C_+^{a_k}})$ over cones $\pm \overset{*}{C_+^{a_k}}$ for almost all $\xi'' = (\xi_1, \cdots, \xi_k)$ with estimates

$$|\sigma_{\neq}^{\pm 1}(x_0, \xi + i\tau)| \leq c_1(1 + |\xi| + |\tau|)^{\pm æ_k},$$

$$|\sigma_{=}^{\pm 1}(x_0, \xi - i\tau)| \leq c_2(1 + |\xi| + |\tau|)^{\pm(\alpha - æ_k)},$$

where c_1, c_2 are constants.

The number $æ_k$ is called an index of the wave factorization.

The following theorem in general was proved in [11]. There are some examples of elliptic symbols admitting the wave factorization.

Let us denote by S_k a smooth sub-manifold of M consisting of k-wedge points.

Theorem 2.7 *If the elliptic local symbol $\sigma(x_0, \xi)$ admits wave factorization with respect to the wedge $W_+^{a,k}$ for all $x_0 \in S_k$ and $|æ - s| < 1/2$ then all local representatives $A_{x_0}, x_0 \in S_k$, are invertible.*

2.6 The Bochner Operator

Let $S(\mathbb{R}^m)$ be the Schwartz space of infinitely differentiable rapidly decreasing at infinity functions. Let us define the following function

$$B(z) = \int\limits_{C_+^a} e^{iy\cdot z} dy, \quad z = x + i\tau \in T(\overset{*}{C_+^a})$$

and introduce the following

Definition 2.8 A Bochner operator is called the following linear operator

$$(Bu)(x) = \lim_{\tau \to 0} \int\limits_{\mathbb{R}^m} B(z - y)u(y)dy,$$

where $\tau \to 0$ along arbitrary non-tangential way, $\tau \in \overset{*}{C_+^a}$ [1, 16].

Remark 2.9 For this case C_+^a the Bochner kernel can be calculated exactly [10, 16]

$$B(z) = \frac{a\Gamma(m/2)}{2\pi^{\frac{m+2}{2}}} \frac{1}{\left(z'^2 - a^2 z_m^2\right)^{m/2}}, \quad z' = (z_1, \cdots, z_{m-1}),$$

where Γ is Euler Γ-function, but all above is valid for arbitrary **sharp convex cone** in $\mathbb{R}^m$ although we don't know an explicit form of the kernel $B(z)$.

Proposition 2.10 *The operator $B : L_2(\mathbb{R}^m) \to L_2(\mathbb{R}^m)$ is a linear bounded operator.*

Proof It follows from the fact that the operator B is Fourier image of a multiplication operator on an indicator of the cone C_+^a. □

Remark 2.11 It is easy to prove that $B : H^s(\mathbb{R}^m) \to H^s(\mathbb{R}^m)$ is also linear bounded operator for $|s| < 1/2$.

Let us note that the Bochner operator plays an important role and permits to construct an inverse operator for local operators (2.3), (2.4) [10, 11].

3 Hidden Parameters

Everywhere above we have assumed that following parameters are constants.

3.1 Order of an Operator

First an order of an pseudo-differential operator can vary from a point to a point. Simple example is an elliptic local symbol of following kind

$$\sigma(x_0, \xi) = (1 + |\xi|^2)^{\alpha(x_0)}.$$

So there is the following

Problem 3.1 What one can say on boundedness and invertibility of such a pseudo-differential operator in Sobolev–Slobodetskii spaces $H^s(M)$?

3.2 Index of Wave factorization

Since index of factorization (according to Vishik–Eskin theory) determines a quantity of boundary conditions and index of wave factorization also it is very interesting situation when such indices vary from a point to a point. Thus the following question arises.

Problem 3.2 Is it possible the situation when one needs different quantity of boundary conditions on distinct parts of a boundary?

3.3 A Variable Size of a Cone

Here we consider a case when size of a cone varies from a point to a point. It means that size of a canonical cone C_+^a can vary, in other words we need to consider a cone of a variable size $C_+^{a(x_0)}$. According to the definition of the operator B one can construct the following operator

$$(B_{var}u)(x) = \frac{a\Gamma(m/2)}{2\pi^{\frac{m+2}{2}}} \lim_{\tau\to 0+} \int\limits_{\mathbb{R}^m} \frac{u(y)dy}{((x'-y')^2 - a^2(x)(x_m - y_m + i\tau)^2)^{m/2}}.$$

Problem 3.3 What one can say on boundedness of the operator B_{var} in Sobolev–Slobodetskii spaces $H^s(\mathbb{R}^m)$?

4 A Wedge with a Variable Size

Here we'll consider more complicated manifold M with so-called k-wedges with a variable size. To study such singularities one can apply the developed technique for describing sufficient invertibility conditions of local operators.

Definition 4.1 k-wedge of a variable size $W_+^{a_k(x_0),k}$ is a smooth sub-manifold $S_k \subset M$ consisting of points x_0 in which a local representative of an operator A has the form

$$u(x) \longmapsto \int\limits_{W_+^{a_k(x_0),k}} \int\limits_{\mathbb{R}^m} \tilde{A}(\cdot, \xi) u(y) e^{i(x-y)\cdot\xi} d\xi dy, \quad x \in W_+^{a_k(x_0),k}, \tag{4.1}$$

where the function $a(x_0)$ is defined on S_k, is continuous, takes its values on an interval (b_k, c_k) and has finite limits in points b_k, c_k.

4.1 Refined Theorem and Sufficient Conditions

Let a manifold M be such that its boundary ∂M includes a smooth part and smooth sub-manifolds S_k which are k-wedges, $k = 0, \cdots, m-2$. A sub-manifold S_{m-1} is a closure of a smooth part of a boundary ∂M. For this piece of a boundary one can use Vishik–Eskin theory [3].

In this section we'll add to local representatives the operator (4.1) and formulate the following

Theorem 4.2 *Let A be a pseudo-differential operator with continuous elliptic local symbol $\sigma(x_0, \xi)$. The operator $A : H^s(M) \to H^{s-\alpha}(M)$ has a Fredholm property iff all local representatives $A_{x_0} : H^s(D) \to H^{s-\alpha}(D)$ are invertible. If the local symbol admits the wave factorization with respect to k-wedge points $x_0 \in S_k$, $|æ_k - s| < 1/2, k = 0, \cdots, m-2$, then all such local representatives $A_{x_0} : H^s(D) \to H^{s-\alpha}(D)$ are invertible.*

4.2 From a Half-Space to a Half-Line: Degenerating Wedge

If $b_1 = 0, c_1 = +\infty$ we have the wedge on Fig. 2. For $b_1 = 0$ we obtain a plane, and for $c_1 = +\infty$ we obtain a half-line. The author has made some attempts to describe such local representatives of an operator A [13, 14] but it is not clear up to now how one can work with such singularities.

5 Exotic Singularities

There are a lot of possibilities to construct another types of singularities combining mentioned above cones, wedges and their modifications. Some variants were presented in [12].

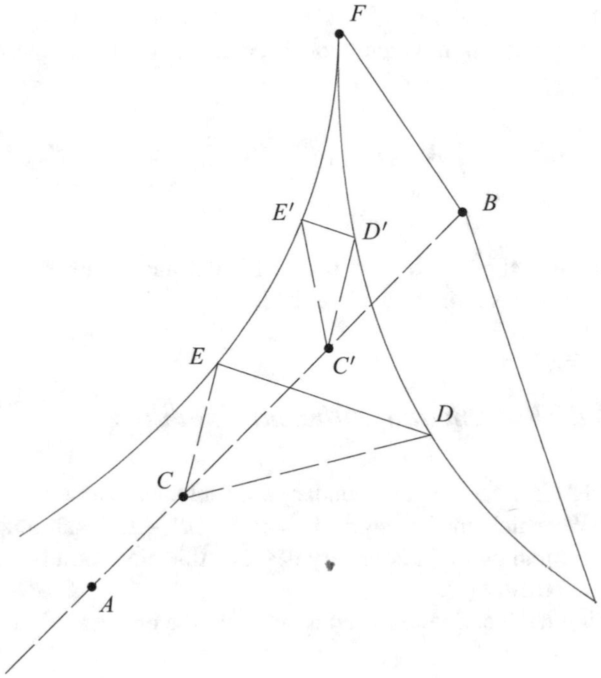

Fig. 2 A wedge of a variable size

6 Conclusion

The author hopes these considerations will help to enlarge a set of admissible manifolds with singular boundaries and to get answers to some formulated questions. It seems for all existing theories of boundary value problems for elliptic pseudo-differential equations on manifolds with singular boundaries one needs an invertibility of local representatives. Finding effective necessary and sufficient conditions for this property is a very hard problem, and any result in this direction will be a great achievement.

Acknowledgements This work was supported by the State contract of the Russian Ministry of Education and Science (contract No 1.7311.2017/B).

References

1. S. Bochner, W.T. Martin, *Several Complex Variables* (Princeton University Press, Princeton, 1948)
2. S. Coriasco, B.-W. Schulze, Edge problems on configurations with model cones of different dimensions. Osaka J. Math. **43**, 63–102 (2006)

3. G. Eskin, *Boundary Value Problems for Elliptic Pseudodifferential Equations* (AMS, Providence, 1981)
4. S. Hofmann, M. Mitrea, M. Taylor, Singular integrals and elliptic boundary problems on regular semmes–Kenig–Toro domains. Int. Math. Res. Not. **14**, 2567–2865 (2010)
5. R.B. Melrose, Pseudodifferential operators, corners and singular limits. In *Proceeding of the International Congress of Mathematicians, 21–29 August 1990, Kyoto, Japan*, ed. by Satake, vol. I (Springer, Tokyo, 1991), pp. 217–234
6. S. Moroianu, V. Nistor, Index and homology of pseudodifferential operators on manifolds with boundary. In *Perspectives in Operator Algebras and Mathematical Physics*. Theta Series in Advanced Mathematics, vol. 8 (Theta, Bucharest, 2008), pp. 123–148
7. V. Nistor, Analysis on singular spaces: lie manifolds and operator algebras. J. Geom. Phys. **105**, 75–101 (2016)
8. M. Ruzhansky, N. Tokmagambetov, Nonharmonic analysis of boundary value problems. Int. Math. Res. Not. **12**, 3548–3615 (2016)
9. I.B. Simonenko, A local method in the theory of translation invariant operators and their envelopings (in Russian) (CVVR Publishing, Rostov on Don, 2007)
10. V.B. Vasil'ev, Regularization of multidimensional singular integral equations in non-smooth domains. Trans. Mosc. Math. Soc. **59**, 65–93 (1998)
11. V.B. Vasil'ev, *Wave Factorization of Elliptic Symbols: Theory and Applications* (Kluwer Academic Publishers, Dordrecht, 2000)
12. V.B. Vasilev, Pseudodifferential equations in cones with conjugate points on the boundary. Differ. Equ. **51**, 1113–1125 (2015)
13. V.B. Vasilyev, Asymptotical analysis of singularities for pseudo differential equations in canonical non-smooth domains. In *Integral Methods in Science and Engineering*, ed. by C. Constanda, P. Harris (Birkhäuser, Boston, 2011), pp. 379–390
14. V.B. Vasilyev, Pseudo differential equations on manifolds with non-smooth boundaries. In *Differential and Difference Equations and Applications*, ed. by S. Pinelas et al. Springer Proceedings in Mathematics & Statistics, vol. 47 (Birkhäuser, Basel, 2013), pp. 625–637
15. V.B. Vasilyev, New constructions in the theory of elliptic boundary value problems. In *Integral Methods in Science and Engineering*, ed. by C. Constanda, A. Kirsch (Birkhäuser, New York, 2015), pp. 629–641
16. V.S. Vladimirov, *Methods of the Theory of Functions of Many Complex Variables* (Dover Publications, Mineola, 2007)

Gravity Driven Flow Past the Bottom with Small Waviness

R. Wojnar and W. Bielski

Abstract We propose an introductory study of gravity driven Stokesian flow past the wavy bottom, based on Adler's et al. papers. In examples the waviness is described by a sinus function and its amplitude is small, up to $O(\varepsilon^2)$. A correction to Hagen-Poiseuille's type free-flow solution is found. A contribution of capillary surface tension is discussed.

Keywords Stokes' equation • Asymptotic and Fourier's expansions • Roughness • Obstacles

Mathematics Subject Classification (2010) Primary 35C20, 35J05; Secondary 41A58, 42B05, 85A30, 86A05

1 Introduction

Alexander Malevich, Vladimir Mityushev and Pierre Adler, [8–10], and recently Roman Czapla, V. Mityushev and Wojciech Nawalaniec, [2], have shown how to apply the asymptotic analysis to reduce the problem of flow in channels with curvilinear walls to the problem of flow with the plane ones. After these authors we study the influence of bottom small waviness on the free flow.

We consider a Stokesian gravity driven free flow past a wavy bottom described by a non-plane interface surface $S(x)$. For simplicity, we are dealing with two dimensional steady problem, described by two position co-ordinates $\boldsymbol{x} = (x, y)$.

R. Wojnar
Institute of Fundamental Technological Research PAS, ul. Pawińskiego 5B, 02-106 Warszawa, Poland
e-mail: rwojnar@ippt.gov.pl

W. Bielski (✉)
Institute of Geophysics, Polish Academy of Sciences, Księcia Janusza 64, 01-452 Warsaw, Poland
e-mail: wbielski@igf.edu.pl

P. Drygaś, S. Rogosin (eds.), *Modern Problems in Applied Analysis*, Trends in Mathematics, https://doi.org/10.1007/978-3-319-72640-3_13

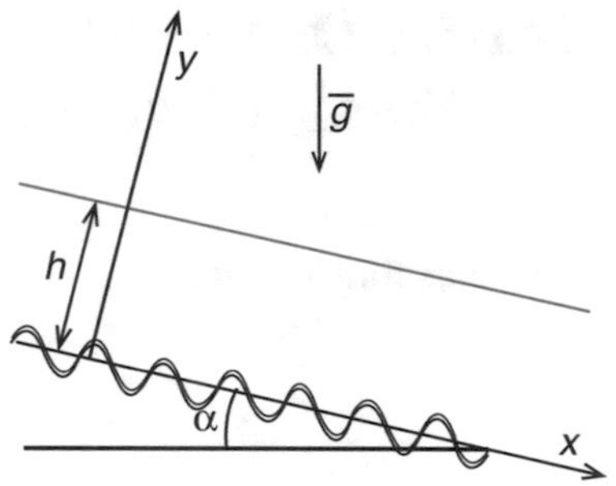

Fig. 1 The gravity driven flow past a wavy bottom with the mean slope α to the horizontal plane. The vector $\boldsymbol{g}$ denotes the gravity acceleration. The depth of the stream in subsequent calculations is taken as $h = 1$. The proportions of the bottom waviness are exaggerated

The task is related to problems of stream flows past the rough bottom, in peculiarity in channels with asperity and obstacles on the bottom.

The issue was treated in numerous experimental and theoretical works, for example [1, 3–6].

1.1 Geometry

The scheme of the problem is given in Fig. 1. It is admitted that the bottom of the channel is not plane but curvilinear. It is, however, plane in the mean. The upper surface of the flow is free. The x-axis is directed parallelly to the bottom descend, and y-axis is perpendicular to the bottom. In an example, we will accept the bottom surface $S(x)$ described by the sinusoid,

$$y = S(x) = \varepsilon\, a\, \cos x$$

where $S(x)$ is a smooth periodic function and ε is a small parameter. The mean value of the $S(x)$ is zero. The coefficient a is a number. For the infinitely differentiable function $S(x)$ a cascade of boundary value problems is deduced. The boundary conditions are substituted by Maclaurin's expansions, and the solution (the velocity and pressure fields) is calculated in the form of both, ε expansions and Fourier's series. The case $\varepsilon = 0$ corresponds to the zeroth approximation problem. The mean value of $S(x)$ is also zero.

1.2 The Problem

Let $\boldsymbol{v} = (u, v)$ be an unknown two dimensional velocity field, and p be an unknown field of the pressure. We have $u = u(x, y)$, $v = v(x, y)$ and $p = p(x, y)$. These fields

satisfy the following set:
equation of incompressibility

$$\frac{\partial u}{\partial x} + \frac{\partial v}{\partial y} = 0 \tag{1.1}$$

and Stokes' equations

$$\begin{aligned} \frac{\partial^2 u}{\partial x^2} + \frac{\partial^2 u}{\partial y^2} - \frac{\partial p}{\partial x} + 1 = 0 \\ \frac{\partial^2 v}{\partial x^2} + \frac{\partial^2 v}{\partial y^2} - \frac{\partial p}{\partial y} - \mathrm{ctg}\alpha = 0 \end{aligned} \tag{1.2}$$

The parameter α is an angle of mean inclination of the bottom to the horizontal plane, (or the bottom inclination at $\varepsilon = 0$), see Fig. 1. The equations are subject to the boundary conditions

$$u = 0 \qquad \text{at } S(x) \tag{1.3}$$

$$p = p_A \quad \text{at the free surface of the stream} \tag{1.4}$$

where p_A is an atmospheric pressure.

The equations (1.1)–(1.4) are written in a natural system of units, described in the Appendix 1.

1.3 Zeroth Approximation

We consider the following set of equations treated a zeroth approximation

$$\begin{aligned} \frac{\partial^2 u_0}{\partial y^2} - \frac{\partial p_0}{\partial x} + 1 = 0 \\ \frac{\partial p_0}{\partial y} + \mathrm{ctg}\,\alpha = 0 \end{aligned} \tag{1.5}$$

which are subject to the boundary conditions

$$u_0 = 0 \quad \text{at} \quad y = 0 \tag{1.6}$$

$$p_0 = p_A \quad \text{at} \quad y = 1 \tag{1.7}$$

In this approximation the velocity vector has only one not vanishing component u_0, which depends on one variable y only. The solution reads

$$u_0 = -\frac{1}{2}y^2 + y \tag{1.8}$$

$$p_0 = p_A + (1 - y)\operatorname{ctg}\alpha \tag{1.9}$$

cf. [7]. Moreover

$$v_0 = 0 \tag{1.10}$$

This zeroth solution can be regarded as Hagen-Poiseuille's solution corresponding to the free-flow.

2 General Solution

Consider the following set of homogeneous equations

$$\begin{aligned} &\frac{\partial u}{\partial x} + \frac{\partial v}{\partial y} = 0 \\ &\frac{\partial^2 u}{\partial x^2} + \frac{\partial^2 u}{\partial y^2} - \frac{\partial p}{\partial x} = 0 \\ &\frac{\partial^2 v}{\partial x^2} + \frac{\partial^2 v}{\partial y^2} - \frac{\partial p}{\partial y} = 0 \end{aligned} \tag{2.1}$$

with the unknown functions u, v and p. They are subject to the boundary conditions

$$u = 0 \quad \text{and} \quad v = 0 \quad \text{on} \quad y = S(x) \tag{2.2}$$

In the accepted frame of reference, see Fig. 1, and in a further exemplary calculation the bottom is described by the function

$$y = S(x) = \varepsilon a \cos x \tag{2.3}$$

The pressure p is equal to the atmospheric pressure at the free upper surface of the stream, $p = p_A$. The mean depth of the stream is equal 1.

2.1 Expansion in ε Series

The unknown velocity components and pressure functions are expanded in the ε series

$$
\begin{aligned}
u(x, y) &= \sum_{m=0}^{\infty} u_m(x, y)\, \varepsilon^m \\
v(x, y) &= \sum_{m=0}^{\infty} v_m(x, y)\, \varepsilon^m \\
p(x, y) &= \sum_{m=0}^{\infty} p_m(x, y)\, \varepsilon^m
\end{aligned}
\tag{2.4}
$$

Next, at the bottom, it is at $y = \varepsilon a \cos x$ a Maclaurin's series of a function, say $g = g(x, y)$, is introduced

$$
g(x, \varepsilon a \cos x) = \sum_{m=0}^{\infty} \varepsilon^m \frac{(\varepsilon a \cos x)^m}{m!} \cdot \left. \frac{\partial^m g}{\partial y^m} \right|_{y=0}
$$

By analogy to this expansion, the bottom boundary conditions for the functions (2.4) are represented by Maclaurin's series

$$
\begin{aligned}
u(x, \varepsilon a \cos x) &= \sum_{m=0}^{\infty} \varepsilon^m \sum_{k=0}^{m} \frac{a^k}{k!} \cos^k x \cdot \left. \frac{\partial^k u_{m-k}}{\partial y^k} \right|_{y=0} \\
v(x, \varepsilon a \cos x) &= \sum_{m=0}^{\infty} \varepsilon^m \sum_{k=0}^{m} \frac{a^k}{k!} \cos^k x \cdot \left. \frac{\partial^k v_{m-k}}{\partial y^k} \right|_{y=0} \\
p(x, \varepsilon a \cos x) &= \sum_{m=0}^{\infty} \varepsilon^m \sum_{k=0}^{m} \frac{a^k}{k!} \cos^k x \cdot \left. \frac{\partial^k p_{m-k}}{\partial y^k} \right|_{y=0}
\end{aligned}
\tag{2.5}
$$

In this manner the search for solution of the set of equations (2.1) with the boundary conditions (2.2) is substituted by solving these equations with conditions (2.5).

2.2 Reduced Problem

Substituting expansions (2.4) into Eqs. (2.1) leads to the equations

$$
\begin{aligned}
&\frac{\partial u_m}{\partial x} + \frac{\partial v_m}{\partial y} = 0 \\
&\frac{\partial^2 u_m}{\partial x^2} + \frac{\partial^2 u_m}{\partial y^2} - \frac{\partial p_m}{\partial x} = 0 \\
&\frac{\partial^2 v_m}{\partial x^2} + \frac{\partial^2 v_m}{\partial y^2} - \frac{\partial p_m}{\partial y} = 0
\end{aligned}
\tag{2.6}
$$

Solving the problem subject to the boundary conditions given at the curvilinear boundary $y = \varepsilon a \cos x$ was reduced to the problem with the straight boundary $y = 0$ but with modified values of boundary conditions. This means that at each step m the set (2.6) must be solved with the modified boundary conditions (2.5).

The solution u_m, v_m and p_m is looked for in the form of Fourier's series

$$\begin{aligned} u_m(x,y) &= \sum_{s=1}^{\infty} \alpha_s^{(m)}(y)\,\frac{\partial}{\partial x}\,(A^{(m)}\,\sin sx + B^{(m)}\,\cos sx) \\ v_m(x,y) &= \sum_{s=1}^{\infty} \beta_s^{(m)}(y)\,(A^{(m)}\,\sin sx + B^{(m)}\,\cos sx) \\ p_m(x,y) &= \sum_{s=1}^{\infty} \gamma_s^{(m)}(y)\,(A^{(m)}\,\sin sx + B^{(m)}\,\cos sx) \end{aligned} \tag{2.7}$$

with y-functions $\alpha_s^{(m)}$, $\beta_s^{(m)}$ and $\gamma_s^{(m)}$ to be found. Also the constants $A^{(m)}$ and $B^{(m)}$ must be evaluated. Notice that $\alpha_s^{(m)}$ is a coefficient of expansion, while α without any subscript is the mean angle of inclination of the bottom, cf. Fig. 1.

Substituting the expansions (2.7) into Eqs. (2.6) we obtain for the incompressibility equation

$$\sum_{s=1}^{\infty}\left(\alpha_s^{(m)}(y)\,s^2 - \frac{\partial}{\partial y}\beta_s^{(m)}(y)\right)(A^{(m)}\sin sx + B^{(m)}\cos sx) = 0$$

and for Stokes' equations

$$\begin{aligned} &\sum_{s=1}^{\infty}\left(\frac{\partial^2}{\partial y^2}\alpha_s^{(m)}(y) - s^2\right)\frac{\partial}{\partial x}(A^{(m)}\sin sx + B^{(m)}\cos sx) = \\ &\qquad\qquad \sum_{s=1}^{\infty}\gamma_s^{(m)}(y)\,\frac{\partial}{\partial x}(A^{(m)}\sin sx + B^{(m)}\cos sx) \\ &\sum_{s=1}^{\infty}\left(\frac{\partial^2}{\partial y^2}\beta_s^{(m)}(y) - s^2\right)(A^{(m)}\,\sin sx + B^{(m)}\,\cos sx) = \\ &\qquad\qquad \sum_{s=1}^{\infty}\frac{\partial}{\partial y}\gamma_s^{(m)}(y)\,(A^{(m)}\sin sx + B^{(m)}\cos sx) \end{aligned} \tag{2.8}$$

Therefore,

$$\frac{\mathrm{d}}{\mathrm{d}y}\,\beta_s^{(m)}(y) - s^2\,\alpha_s^{(m)}(y) = 0 \tag{2.9}$$

$$\frac{\mathrm{d}^2}{\mathrm{d}y^2}\alpha_s^{(m)}(y) - s^2\alpha_s^{(m)}(y) - \gamma_s^{(m)}(y) = 0 \tag{2.10}$$

$$\frac{\mathrm{d}^2}{\mathrm{d}y^2}\beta_s^{(m)}(y) - s^2\beta_s^{(m)}(y) - \frac{\mathrm{d}}{\mathrm{d}y}\,\gamma_s^{(m)}(y) = 0 \tag{2.11}$$

Since the representation applied in the expansions (2.7) is a special case of the trig-function introduced in [8], the system of ordinary differential equations (2.9)–(2.11) is the same as that derived and discussed in [8].

2.3 Solution of the System

After [8], to solve the system of equations (2.9)–(2.11) we differentiate Eq. (2.10)

$$\frac{\mathrm{d}^3}{\mathrm{d}y^3}\alpha_s^{(m)}(y) - s^2\,\frac{\mathrm{d}}{\mathrm{d}y}\alpha_s^{(m)}(y) - \frac{\mathrm{d}}{\mathrm{d}y}\gamma_s^{(m)}(y) = 0$$

and subtract (2.11) from the result

$$\frac{\mathrm{d}^3}{\mathrm{d}y^3}\alpha_s^{(m)}(y) - s^2\,\frac{\mathrm{d}}{\mathrm{d}y}\alpha_s^{(m)}(y) - \frac{\mathrm{d}^2}{\mathrm{d}y^2}\beta_s^{(m)}(y) + s^2\beta_s^{(m)}(y) = 0$$

In this equation we substitute the function $\alpha_s^{(m)}$ expressed by Eq. (2.9)

$$\alpha_s^{(m)}(y) = \frac{1}{s^2}\,\frac{\mathrm{d}}{\mathrm{d}y}\,\beta_s^{(m)}(y)$$

and get

$$\frac{1}{s^2}\,\frac{\mathrm{d}^4}{\mathrm{d}y^4}\beta_s^{(m)}(y) - 2\,\frac{\mathrm{d}^2}{\mathrm{d}y^2}\beta_s^{(m)}(y) + s^2\,\beta_s^{(m)} = 0$$

The discriminant of the last equation is zero, and the general solution reads

$$\beta_s^{(m)}(y) = (C_1^{(sm)}y + C_2^{(sm)})\,\mathrm{e}^{sy} + (C_3^{(sm)}y + C_4^{(sm)})\,\mathrm{e}^{-sy} \tag{2.12}$$

Simultaneously, by (2.9) and (2.10) we obtain

$$\begin{aligned}\alpha_s^{(m)}(y) = \frac{1}{s^2}\Big\{&[C_1^{(sm)} + (C_1^{(sm)}y + C_2^{(sm)})\,s]\,\mathrm{e}^{sy} + \\ &+ [C_3^{(sm)} - (C_3^{(sm)}y + C_4^{(sm)})\,s]\,\mathrm{e}^{-sy}\Big\}\end{aligned} \tag{2.13}$$

and

$$\gamma_s^{(m)}(y) = 2C_1^{(sm)}\,\mathrm{e}^{sy} + 2C_3^{(sm)}\,\mathrm{e}^{-sy} \tag{2.14}$$

Three last expressions, when substituted successively into Eqs. (2.7) and (2.4) would give the desired solution. Before, however, the boundary conditions should be accounted for determining constants $C_i^{(sm)}$, $i = 1, 2, 3, 4$.

3 An Example

Without loss of generality we put

$$a = 1 \tag{3.1}$$

In approximation up to $O(\varepsilon^2)$ we have

$$\begin{aligned} u(x, y) &= u_0(x, y) + \varepsilon u_1(x, y) \\ v(x, y) &= v_0(x, y) + \varepsilon v_1(x, y) \\ p(x, y) &= p_0(x, y) + \varepsilon p_1(x, y) \end{aligned} \tag{3.2}$$

with the boundary conditions at $y = 1$

$$\begin{aligned} v(x, 1) &= 0 \\ p(x, 1) &= p_{\mathrm{A}} \end{aligned} \tag{3.3}$$

The following equations should be satisfied

$$\begin{aligned} &\frac{\partial}{\partial x}[u_0(x, y) + \varepsilon u_1(x, y)]) + \frac{\partial}{\partial y}[v_0(x, y) + \varepsilon v_1(x, y)] = 0 \\ &\frac{\partial^2}{\partial x^2}[u_0(x, y) + \varepsilon u_1(x, y)] + \frac{\partial^2}{\partial y^2}[u_0(x, y) + \varepsilon u_1(x, y)] - \\ &\qquad \frac{\partial}{\partial x}[p_0(x, y) + \varepsilon p_1(x, y)] + 1 = 0 \\ &\frac{\partial^2}{\partial x^2}[v_0(x, y) + \varepsilon v_1(x, y)] + \frac{\partial^2}{\partial y^2}[v_0(x, y) + \varepsilon v_1(x, y)]) - \\ &\qquad \frac{\partial}{\partial y}[p_0(x, y) + \varepsilon p_1(x, y)] + \operatorname{ctg} \alpha = 0 \end{aligned}$$

or

$$\begin{aligned} &\frac{\partial}{\partial x}u_0(x, y) + \frac{\partial}{\partial y}v_0(x, y) + \varepsilon\left(\frac{\partial}{\partial x}u_1(x, y) + \frac{\partial}{\partial y}v_1(x, y)\right) = 0 \\ &1 + \left(\frac{\partial^2}{\partial x^2} + \frac{\partial^2}{\partial y^2}\right)u_0(x, y) - \frac{\partial}{\partial x}p_0(x, y) + \\ &\quad \varepsilon\left\{\left(\frac{\partial^2}{\partial x^2} + \frac{\partial^2}{\partial y^2}\right)u_1(x, y) - \frac{\partial}{\partial x}p_1(x, y)\right\} = 0 \\ &\operatorname{ctg}\alpha + \left(\frac{\partial^2}{\partial x^2} + \frac{\partial^2}{\partial y^2}\right)v_0(x, y) - \frac{\partial}{\partial x}p_0(x, y) + \\ &\quad \varepsilon\left\{\left(\frac{\partial^2}{\partial x^2} + \frac{\partial^2}{\partial y^2}\right)v_1(x, y) - \frac{\partial}{\partial x}p_1(x, y)\right\} = 0 \end{aligned} \tag{3.4}$$

3.1 Two Problems: ε^0 and ε^1

Now, we consider coefficient at ε^0, it is:

Problem ε^0, which gives, see Sect. 1.3,

$$u_0 = y - \frac{1}{2}y^2, \quad v_0 = 0, \quad \text{and} \quad p_0 = p_A + (1-y)\operatorname{ctg}\alpha \tag{3.5}$$

and Problem ε^1

$$\begin{aligned}
&\frac{\partial}{\partial x}u_1(x,y) + \frac{\partial}{\partial y}v_1(x,y) = 0\\
&\left(\frac{\partial^2}{\partial x^2} + \frac{\partial^2}{\partial y^2}\right)u_1(x,y) - \frac{\partial}{\partial x}p_1(x,y) = 0\\
&\left(\frac{\partial^2}{\partial x^2} + \frac{\partial^2}{\partial y^2}\right)v_1(x,y) - \frac{\partial}{\partial x}p_1(x,y) = 0
\end{aligned} \tag{3.6}$$

To find the first approximations u_1, v_1 and p_1 in the form according to [8] one puts

$$\begin{aligned}
u_1(x,y) &= u_{11}(y)\sin x + u_{12}(y)\cos x\\
v_1(x,y) &= v_{11}(y)\sin x + v_{12}(y)\cos x\\
p_1(x,y) &= p_{11}(y)\sin x + p_{12}(y)\cos x
\end{aligned} \tag{3.7}$$

cf. the Appendix 2, Eqs. (A.12). By relations (2.12)–(2.14)

$$\alpha_1^{(1)}(y) = [C_1 + (C_1 y + C_2)]\,e^y + [C_3 - (C_3 y + C_4]\,e^{-y} \tag{3.8}$$

$$\beta_1^{(1)}(y) = (C_1 y + C_2)\,e^y + (C_3 y + C_4)\,e^{-y} \tag{3.9}$$

and

$$\gamma_1^{(1)}(y) = 2C_1\,e^y + 2C_3\,e^{-y} \tag{3.10}$$

For simplifying notation, we omitted the superscripts everywhere over the constants C_1, C_2, C_3 and C_4, while the subscripts and coefficients at x are put $s = 1$.

3.2 Solution in the First Approximations

The constants C_1, C_2, C_3 and C_4 are found in the Appendix 2. Finally, we get

$$
\begin{aligned}
u(x,y) &= u_0(x,y) + \varepsilon u_1(x,y) = \\
&= y - \frac{1}{2}y^2 + \varepsilon \frac{-e^2 + e^{2y} + \left(e^2 + e^{2y}\right) y}{-1 + e^2} e^{-y} \cos x \\
v(x,y) &= v_0(x,y) + \varepsilon v_1(x,y) = \varepsilon \frac{-e^2 + e^{2y}}{-1 + e^2} e^{-y} y \sin x \\
p(x,y) &= p_0(x,y) + \varepsilon p_1(x,y) = \\
&= p_A + (1-y)\mathrm{ctg}\alpha + \varepsilon\, 2 \frac{-e^2 + e^{2y}}{-1 + e^2} e^{-y} \sin x
\end{aligned}
\tag{3.11}
$$

From the last set of results we are learning that the ε - corrections to the transversal velocity component correction v_1 and the pressure correction p_1 are opposite in the phase to the longitudinal velocity component correction u_1. One observes that a higher velocity is accompanied by a lower pressure, and one can look for a far analogy to Bernoulli's law.

3.3 Results

Figure 2 shows the dependence of the longitudinal velocity u on the transversal co-ordinate y are shown, and Fig. 3 is illustrating the analogical dependence of the transversal velocity component v, for different values of the parameter ε. In our notation, since $a = 1$, the parameter ε denotes the amplitude of the bottom waviness. Since the first order correction u_1 changes its sign on the y axis the maximum value of the velocity u changes its position up and down on the x axis according to variations of *cosx* function. Also the transversal velocity v changes its values periodically along the x axis.

What concerns the pressure correction

$$
p_1 = 2 \frac{-e^2 + e^{2y}}{-1 + e^2} e^{-y} \sin x
$$

we observe that on the bottom ($y = 0$) it varies from (-2) to (2) dependent on value of the function $\sin x$, and on the upper surface ($y = 1$) it has always the value zero. It is very small in comparison with the atmospheric pressure which in natural units equals 10^6, cf. the Appendix 1. However, it is not small with comparison the surface tension at the water-air interface which equals $\sigma = 7.25$, cf. the Appendix 5.

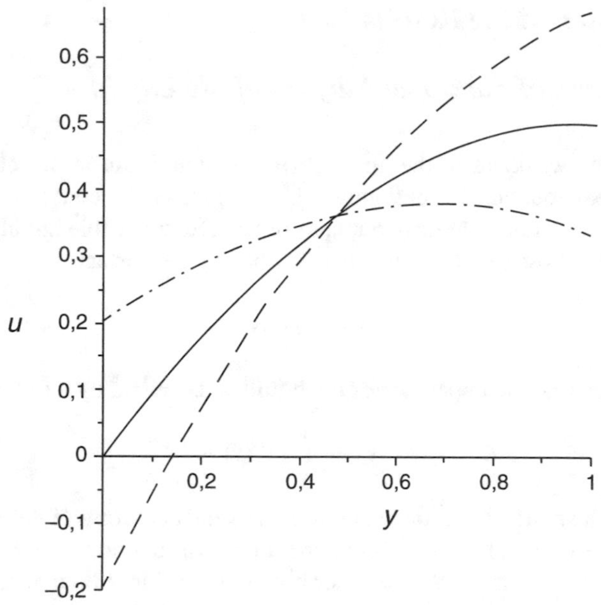

Fig. 2 The longitudinal velocity component u versus the transversal co-ordinate y, for three different values of ε, it is for $\varepsilon = 0$ (solid), $\varepsilon = 0.2$ (dash) and $\varepsilon = -0.2$ (dashdot). The sign minus corresponds to values $\cos x = -1$

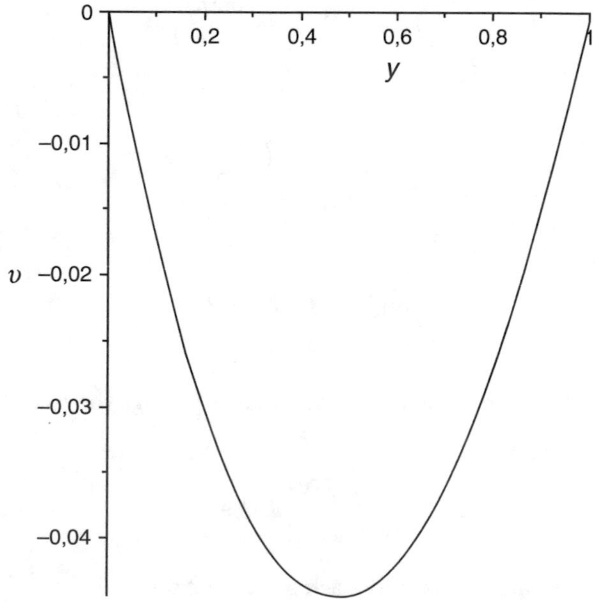

Fig. 3 The transversal velocity component v versus the transversal co-ordinate y for $\varepsilon = 0.2$

4 Discussion of the Results

4.1 Waviness of the Upper Surface of the Liquid

To estimate the waviness of the upper surface of the liquid in our channel, let us write the global equation of continuity, which expresses the conservation of matter (or the constancy of its volume in our case of an incompressible liquid). In example of Sect. 3, the bottom of the channel is described by the surface

$$y_{\mathrm{B}} = \varepsilon \cos x$$

We can assume that the upper side of the liquid is described by a function

$$y_{\mathrm{A}} = 1 + \zeta(x)$$

where $\zeta(x)$ is a small correction which describes the departure of the liquid surface from the plane $y = 1$. It is to estimate the unknown function $\zeta = \zeta(x)$ we built a global continuity equation. We consider a volume of liquid bounded by two plane cross-sections of the channel at a distance $\mathrm{d}x$ apart. In unit time a volume $(Su)_x$ of liquid flows through one plane, and a volume $(Su)_{x+\mathrm{d}x}$ through the other. The volume of liquid between the two planes changes by

$$(Su)_{x+\mathrm{d}x} - (Su)_x = \frac{\partial (Su)_x}{\partial x}\,\mathrm{d}x$$

But

$$S(x) = 1 \cdot \int_{y_{\mathrm{B}}}^{y_{\mathrm{A}}} \mathrm{d}y$$

and

$$\frac{\partial (Su)_x}{\partial x} = \frac{\partial}{\partial x} \int_{\varepsilon \cos x}^{1+\zeta(x)} u\,\mathrm{d}y$$

The volume of incompressible liquid which flows through the channel is constant. Therefore

$$\frac{\partial}{\partial x} \int_{\varepsilon \cos x}^{1+\zeta(x)} u\,\mathrm{d}y = 0$$

with $u = u_0 + \varepsilon u_1$, and we have for each power of ε respectively

$$\frac{\partial}{\partial x} \int_0^1 u_0\,\mathrm{d}y = 0 \quad \text{and} \quad \frac{\partial}{\partial x} \cdot \int_{\varepsilon \cos x}^{1+\zeta(x)} u_1\,\mathrm{d}y = 0$$

The first equation of the last two is satisfied identically, and the second gives

$$\frac{\partial}{\partial x}\int_{\varepsilon\cos x}^{1+\zeta(x)}\{-\mathrm{e}^2\mathrm{e}^{-y}+\mathrm{e}^y+\mathrm{e}^2\,\mathrm{e}^{-y}y+\mathrm{e}^y y\}\cos x\,\mathrm{d}y=0 \tag{4.1}$$

where the result of Sect. 3

$$u_1(x,y)=\frac{-\mathrm{e}^2+\mathrm{e}^{2y}+(\mathrm{e}^2+\mathrm{e}^{2y})\,y}{\mathrm{e}^2-1}\,\mathrm{e}^{-y}\,\cos x$$

was used.

Our problem, as described by Stokes' equation is a linear one. We remain that linear causality describes a relationship of proportionality between a given cause and a given effect that stays constant over time. Small causes always produce small effects; large causes always produce large effects. Since linear interactions exclude feedback, the intensity of the effect will tend to be proportional to that of the cause. The magnitude of an effect is proportional to the magnitude of its cause, cf. http://complexitylabs.io/linear-causality/. Hence, we admit that both quantities, ε and ζ are of the same order. As the curvature of any curve $f(x)$ is of the order $|\mathrm{d}^2f(x)/\mathrm{d}x^2|$, the maximal values of the surface curvature are of the order $\varepsilon\cos 0=\varepsilon$.

4.2 *Surface Phenomena*

It is well known that capillary gravity waves can arise in the presence the gravitational field at the liquid surface, [7]. They are described by the dispersion relation

$$\omega^2=gk+\sigma k^3/\rho$$

where ω is the circular frequency of the wave, k is the wave number, with $\lambda=2\pi/k$ being the wavelength, and h is the depth of the liquid, see Fig. 1.

In our estimation we accept that the waviness of the upper surface is comparable with the waviness of the bottom. This waviness in metrical system is described by $\cos(x/h)$. The corresponding curvature is

$$\frac{\mathrm{d}^2\cos(x/h)}{\mathrm{d}x^2}=-\frac{1}{h^2}\cos(x/h)$$

For thin films, with h sufficiently small, the contribution of surface tension cannot be neglected, if one looks for the appropriate physical description of the gravity driven flow, cf. the Appendix 5.

Equations used in our paper, the continuity equation (1.1) and Stokes' equations (1.2) are time independent. Moreover, Eqs. (1.2) are lacking in the inertial term. Thus, these equations cannot describe any form undulatory motion, and neither gravity neither capillary waves come within the scope of this formulation. However, in principle the phenomenon of surface tension can affect our results. In principle, to obtain an integral description of surface phenomena the complete set of Navier-Stokes' equations should be accounted for.

5 Remarks

We have presented a study of a two-dimensional flow past a wavy impermeable bottom in the first order approximation of the parameter ε, which is a measure of the waviness amplitude. In this paper example the bottom waviness was accepted in a simple sinusoidal form.

The bottom waviness results in appearance of a transversal velocity, and it is a new qualitative feature of the flow in comparison with the flow past the plane bottom. Also the longitudinal flow velocity is subduing periodical changes.

We hope that our results will be useful in practical applications. Wavy bottom surfaces are found in circumstances related to deformable or granular media, for example wind-wave interactions with the surface of the sand desert, of the ocean or sand dunes in the river or sea bottom. The influence of bottom plants on the flow in rivers is crucial in geophysical and agricultural problems, where different obstacles at a river bottom may appear, as stones or plants, [6].

What concerns scientific meaning, it is relevant to explain mechanisms involved in complex phenomena observed in the nature, such as the formation of dunes or ripples in the deserts and in the ocean beds, cf. [3, 12].

Acknowledgements The authors gratefully acknowledge many helpful suggestions from Professor Vladimir V. Mityushev.

This work was partially supported within statutory activities No 3841/E-41/S/2017 of the Ministry of Science and Higher Education of Poland.

Appendix 1: Natural Units for the Problem

Let the module of velocity of the main stream can be characterized by the quantity u, and the geometrical properties of the flow can be denoted by l. Then any flow is specified by three parameters u, l and the viscosity η. The only one dimensionless quantity which can be formed from the above three is called Reynolds' number

$$R \equiv \frac{\rho}{\eta} ul \tag{A.1}$$

Navier-Stokes' equation is considerably simplified in the case of flow with small Reynolds number R. For steady flow of an incompressible fluid with constant viscosity η this equation reads

$$(\boldsymbol{v} \cdot \nabla)\boldsymbol{v} = -\frac{1}{\rho}\nabla p + \boldsymbol{b} + \frac{\eta}{\rho}\Delta\boldsymbol{v} \tag{A.2}$$

The term $(\boldsymbol{v} \cdot \nabla)\boldsymbol{v}$ is of the order of magnitude of (u^2/l), and the quantity $(\eta/\rho)\,\Delta\boldsymbol{v}$ is of the order of $\eta u/(\rho l^2)$. The ratio of two is just Reynolds' number R. Hence the term $(\boldsymbol{v} \cdot \nabla)\boldsymbol{v}$ may be neglected for small R. Then the equation of motion reduces to a linear equation

$$\eta\,\Delta\boldsymbol{v} - \nabla p + \boldsymbol{b} = 0 \tag{A.3}$$

known as Stokes' equation, which together with the equation of continuity (1) and appropriate boundary conditions completely determines the motion, [7].

We notice yet that the smallness of R is not necessary to linearize Stokes' equation in the case of laminar flow with the velocity $\boldsymbol{v} = (v, 0, 0)$, if v does not depend on x_1. In this case the nonlinear term at the left hand side of Eq. (A.2) identically vanishes.

In analogy with Reynolds' number R, we take as a natural for this problem the depth h of the fluid as the unit of length $\breve{x}$. As the unit of velocity $\breve{v}$ we accept three times repeated the mean velocity $3 \cdot v^{\text{mean}}$. The natural unit of pressure $\breve{p}$ is the x component of the hydrostatic pressure at the bottom, what assures the unit value of the x component of the body force. Thus, cf. Fig. 1,

$$\begin{aligned} \breve{x} &= h \\ \breve{v} &= 3 \cdot v^{\text{mean}} = \frac{\rho}{\eta} h^2 g \sin\alpha \\ \breve{p} &= \frac{\eta}{h} 3\, v^{\text{mean}} = \rho h g \sin\alpha \end{aligned} \tag{A.4}$$

Consequently, the unit of time is

$$\breve{t} = \frac{\breve{x}}{\breve{v}} = \frac{h}{3 \cdot v^{\text{mean}}} = \frac{\eta}{\rho h g \sin\alpha} \tag{A.5}$$

Then new variables $\widetilde{x}$, $\widetilde{v}$ and $\widetilde{p}$ are introduced:

$$\widetilde{x} \equiv \frac{x}{\breve{x}}, \quad \widetilde{v} \equiv \frac{v}{\breve{v}} \quad \text{and} \quad \widetilde{p} \equiv \frac{p}{\breve{p}} \tag{A.6}$$

In the new variables the zeroth approximation of Stokes' equation has the form

$$\frac{\partial^2 \widetilde{v}}{\partial \widetilde{y}^2} - \frac{\partial \widetilde{p}}{\partial \widetilde{x}} + 1 = 0 \qquad \text{and} \qquad \frac{\partial \widetilde{p}}{\partial \widetilde{y}} + \text{ctg}\alpha = 0$$

cf. Eqs. (1.5).

We take into consideration a typical laboratory experiment with $h = 10\,\mathrm{cm}$. If we approximate the gravity acceleration g as $1000\,\mathrm{cm/s^2}$, then for the water with the density $\rho = 1\,\mathrm{g/cm^3}$ the relation between the natural units and the SI units is

$$\begin{aligned}
\breve{x} &= 10\,\mathrm{cm} \\
\breve{v} &= \frac{1\,\frac{\mathrm{g}}{\mathrm{cm^3}}}{0.01\,\frac{\mathrm{g}}{\mathrm{cm\ s}}}\cdot\frac{10^2\,\mathrm{cm}^2}{3}\cdot 1000\frac{\mathrm{cm}}{\mathrm{s}^2}\cdot\sin\alpha = \frac{10}{3}\cdot 10^6\cdot\frac{\mathrm{cm}}{\mathrm{s}}\cdot\sin\alpha \\
\breve{p} &= 1\frac{\mathrm{g}}{\mathrm{cm}^3}\cdot 10\,\mathrm{cm}\cdot 1000\frac{\mathrm{cm}}{\mathrm{s}^2}\cdot\sin\alpha = 10^4\frac{\mathrm{g\ cm}}{\mathrm{s}^2}\cdot\frac{1}{\mathrm{cm}^2}\cdot\sin\alpha \\
\breve{t} &= \frac{0.01\cdot\frac{\mathrm{g}}{\mathrm{cm\ s}}}{1\frac{\mathrm{g}}{\mathrm{cm}^3}\cdot 10\,\mathrm{cm}\cdot 1000\frac{\mathrm{g\ cm}}{\mathrm{s}^2}\cdot\sin\alpha} = \frac{10^{-6}\mathrm{s}}{\sin\alpha}
\end{aligned} \tag{A.7}$$

In peculiar, we have also

$$1\cdot\frac{\mathrm{g\ cm}}{\mathrm{s}^2}\cdot\frac{1}{\mathrm{cm}^2} = \breve{p}\cdot\frac{10^{-4}}{\sin\alpha}$$

and

$$1s = \breve{t}\cdot 10^6\cdot\sin\alpha$$

Then, the viscosity of the water in these new units is equal to one

$$\eta^{\mathrm{H_2O}} = 0.01\cdot\frac{\mathrm{g}}{\mathrm{cm\ s}} = 0.01\cdot\frac{1}{\mathrm{cm}^2}\frac{\mathrm{g\ cm}}{\mathrm{s}^2}\cdot s = 1\cdot\breve{p}\cdot\breve{t} \tag{A.8}$$

In our example we accept

$$\alpha = \frac{100\,\mathrm{m}}{1000\,\mathrm{km}} = 10^{-4} \tag{A.9}$$

what is an order of the mean slope of the great world rivers profile, cf. the Appendix 3. Then

$$\begin{aligned}
\breve{x} &= 10\,\mathrm{cm} \\
\breve{v} &= 10\cdot 10^2\cdot\frac{\mathrm{cm}}{\mathrm{s}} = 10\cdot\frac{\mathrm{m}}{\mathrm{s}} \\
\breve{p} &= 1\cdot\frac{\mathrm{g\ cm}}{\mathrm{s}^2}\cdot\frac{1}{\mathrm{cm}^2} \\
\breve{t} &= 0.01\,\mathrm{s}
\end{aligned} \tag{A.10}$$

In these units, we find

$$\breve{x}\cdot\frac{\breve{p}}{\breve{v}} = \frac{1}{100}\cdot\frac{\mathrm{g}}{\mathrm{cm\ s}}$$

or

$$1\,\mathrm{P} \equiv 1 \cdot \frac{\mathrm{g}}{\mathrm{cm\ s}} = 100 \cdot \breve{x} \cdot \frac{\breve{p}}{\breve{v}}$$

It is conspicuous for this angle of slope the smallness of the pressure unit. Note that $\breve{p} = 1\,\mathrm{dyn/cm^2} = 0.1\,\mathrm{Pa} \approx 10^{-6}\,\mathrm{atm}$.

As it was mentioned above, the viscosity of water at 20°C is almost exactly 1 cP, and equals 1 in our natural units, cf. Eq. (A.8).

Appendix 2: Elaboration of the Example

General Solution

By equation (2.7) for $m = 1$ we have

$$\begin{aligned} u_1(x, y) &= \alpha_1(y)\,(A\,\cos x - B\,\sin x) \\ v_1(x, y) &= \beta_1(y)\,(A\,\sin x + B\,\cos x) \\ p_1(x, y) &= \gamma_1(y)\,(A\,\sin x + B\,\cos x) \end{aligned} \tag{A.11}$$

where α_1, β_1, and γ_1 are given by Eqs. (3.8)–(3.10) and the superscripts 1 were omitted. After Eqs. (3.7) we find

$$\begin{aligned} u_{11} &= -\{[C_1 + (C_1 y + C_2)]\,\mathrm{e}^y + [C_3 - (C_3 y + C_4]\,\mathrm{e}^{-y}\}\,B \\ u_{12} &= \{[C_1 + (C_1 y + C_2)]\,\mathrm{e}^y + [C_3 - (C_3 y + C_4]\,\mathrm{e}^{-y}\}\,A \\ v_{11} &= \{(C_1 y + C_2)\,\mathrm{e}^y + (C_3 y + C_4)\,\mathrm{e}^{-y}\}\,A \\ v_{12} &= \{(C_1 y + C_2)\,\mathrm{e}^y + (C_3 y + C_4)\,\mathrm{e}^{-y}\}\,B \\ p_{11} &= (2C_1\,\mathrm{e}^y + 2C_3\,\mathrm{e}^{-y})A, \qquad p_{12} = (2C_1\,\mathrm{e}^y + 2C_3\,\mathrm{e}^{-y})B \end{aligned} \tag{A.12}$$

Boundary Conditions for the Example

If the expansions (2.4) are limited up to $O(\varepsilon^2)$, only the terms with $m = 0, 1$ must be left, and the boundary conditions (2.5) reduce to the following ones

$$\begin{aligned} u(x, \varepsilon a \cos x) &\doteq u_0(x, 0) + \varepsilon \left(u_1(x, 0) + a \cos x \cdot \left.\frac{\partial u_0(x, y)}{\partial y}\right|_{y=0} \right) \\ v(x, \varepsilon a \cos x) &\doteq v_0(x, 0) + \varepsilon \left(v_1(x, 0) + a \cos x \cdot \left.\frac{\partial v_0(x, y)}{\partial y}\right|_{y=0} \right) \\ p(x, \varepsilon a \cos x) &\doteq p_0(x, 0) + \varepsilon \left(p_1(x, 0) + a \cos x \cdot \left.\frac{\partial p_0(x, y)}{\partial y}\right|_{y=0} \right) \end{aligned} \tag{A.13}$$

where $\doteq$ means the asymptotic equality with the accuracy $O(\varepsilon^2)$. This symbol was applied in [2].

Next, we submit Fourier's series (2.7) into Maclaurin's expansions (2.5), taken at the bottom boundary up to $O(\varepsilon^2)$ approximation, and receive

$$\begin{aligned}
u(x, \varepsilon\, a\, \cos x) &\doteq u_0(x,0) + \\
&\varepsilon \left(\alpha_1(y)|_{y=0} \frac{\partial}{\partial x} (A \sin x + B \cos x) \right) \\
v(x, \varepsilon\, a\, \cos x) &\doteq v_0(x,0) + \\
&\varepsilon \left(\beta_1^{(1)}(y)\Big|_{y=0} (A \sin x + B \cos x) \right) \\
p(x, \varepsilon\, a\, \cos x) &\doteq p_0(x,0) + \\
&\varepsilon \left(\gamma_1^{(1)}(y)\Big|_{y=0} (A \sin x + B \cos x) \right)
\end{aligned} \tag{A.14}$$

The subscripts and coefficients at x are put $s = 1$. Above, $u_0(x,0)$ and $v_0(x,0)$ are known and equal to zero, while $p_0(x,0)$ must be found from a boundary problem at the free surface of the flowing stream.

From the solutions (2.12)–(2.13) we get

$$\begin{aligned}
\alpha_1^{(1)}(y)\Big|_{y=0} &= C_1 + C_2 + C_3 - C_4 \\
\beta_1^{(1)}(y)\Big|_{y=0} &= C_2 + C_4 \\
\frac{\partial}{\partial y}\beta_1^{(1)}(y)\Big|_{y=0} &= C_1 + C_2 + C_3 - C_4 \\
\gamma_1^{(1)}(y)\Big|_{y=0} &= 2\,(C_1 + C_3) \\
\frac{\partial}{\partial y}\gamma_1^{(1)}(y)\Big|_{y=0} &= 2\,C_1 - 2\,C_3
\end{aligned} \tag{A.15}$$

Here also the superscripts at the coefficients C_1, C_2, C_3 and C_4 were omitted. Substituting (A.13) into (A.12) gives

$$\begin{aligned}
u(x, \varepsilon\, a\, \cos x) &\doteq \\
&\varepsilon\, \{(C_1 + C_2 + C_3 - C_4)(A \cos x - B \sin x) + \cos x\} \\
v(x, \varepsilon\, a\, \cos x) &\doteq \varepsilon\,(C_2 + C_4)\,(A \sin x + B \cos x) \\
p(x, \varepsilon\, a\, \cos x) &\doteq p_0(x,0) + \varepsilon\, 2\,(C_1 + C_3)\,(A \sin x + B \cos x)
\end{aligned} \tag{A.16}$$

Finding Constants

At the bottom

$$u(x, \varepsilon\, a\, \cos x) = 0 \tag{A.17}$$

and

$$v(x, \varepsilon\, a\, \cos x) = 0 \tag{A.18}$$

where both $u(x, \varepsilon\, a\, \cos x)$ and $v(x, \varepsilon\, a\, \cos x)$ are given by (A.16). Also, in good approximation (if surface undulations were neglected)

$$p(x, \varepsilon\, a\, \cos x) = p_A + \operatorname{ctg} \alpha \tag{A.19}$$

These conditions should be satisfied separately at each power of ε. We apply the integral relations

$$\int_{-\pi}^{\pi} \cos kx \sin lx \mathrm{d}x = 0 \quad \text{and} \quad \int_{-\pi}^{\pi} \cos kx \cos lx \mathrm{d}x = \pi\, \delta_{kl}$$

and successively obtain.

From Eqs. (A.14) and (A.15)

$$(C_1 + C_2 + C_3 - C_4) A = -1 \tag{A.20}$$

and

$$B = 0 \tag{A.21}$$

From Eqs. (A.14) and (A.16)

$$C_2 + C_4 = 0 \tag{A.22}$$

After the set (3.2) and Eq. (1.9) we have

$$p = p_A + (1 - y) \operatorname{ctg}\alpha + 2\varepsilon(C_1 \mathrm{e}^y + C_3 \mathrm{e}^{-y}) A \sin x \tag{A.23}$$

To assure the value $p = p_A$ at $y = 1$ we should have

$$C_1 \mathrm{e} + C_3 \mathrm{e}^{-1} = 0 \tag{A.24}$$

If we assume that transversal velocity component v vanishes at the stream surface ($y = 1$)

$$(C_1 + C_2)\mathrm{e} + (C_3 + C_4)\mathrm{e}^{-1} = 0 \tag{A.25}$$

Comparing (A.22), (A.24) and (A.25) we observed that

$$C_2 = 0 = C_4 \tag{A.26}$$

Moreover,

$$AC_1 = \frac{e^2}{e^2 - 1} \qquad \text{and} \quad AC_3 = \frac{1}{e^2 - 1} \tag{A.27}$$

Appendix 3: Slopes of the River Beds: Typical Examples

What concerns the lower Nile river, the city Aswan is elevated 194 m above the sea level, and from this town to the Mediterranean sea the river flows about 1200 km yet.

What concerns the lower Amazon river, the city Manaus is on 92 m elevation and 1500 km away from the Atlantic ocean. In approximation the mean slope of the Amazon profile is

$$\alpha \approx \frac{60\,\text{m}}{1500\,\text{km}} = 0.4 \times 10^{-4}$$

The similar numbers are obtained for other rivers such as Volga, Oka, Mississippi and Wisła (Vistula) below Warsaw. The mean slope of the Wisłok river below Rzeszów until the mouth in Dębno is about 14 m / 60 km $\approx 2 \times 10^{-4}$.

For comparison, the slope of the bridge *Pont du Gard*, which descends 2.5 cm in 456 m has the approximate value of the gradient 0.5×10^{-4}. The bridge is part of the Nîmes aqueduct. The aqueduct was built probably around the reign of the emperor Claudius (41–54 AD). It required constant maintenance by *circitores*, workers responsible for the aqueduct's upkeep, who crawled along the conduit scrubbing the walls clean and removing any vegetation, cf. [11].

Appendix 4: Estimation of the Upper Waviness

After the formula for differentiation of a definite integral whose limits are functions of the differential variable, Eq. (4.1) can be written in the form

$$I_1 + I_2 + I_3 = 0 \tag{A.28}$$

Here

$$\begin{aligned}
I_1 &= -\int_{\varepsilon \cos x}^{1+\zeta(x)} \{-e^2 e^{-y} + e^y + e^2 e^{-y} y + e^y y\} \sin x \, dy \\
I_2 &= \frac{\partial \zeta}{\partial x} \{-e^2 e^{-(1+\zeta)} + e^{1+\zeta} + e^2 e^{-(1+\zeta)}(1+\zeta) + e^{1+\zeta}(1+\zeta)\} \cos x \\
I_3 &= -\varepsilon \sin x \{-e^2 e^{-\varepsilon \cos x} + e^{\varepsilon \cos x} + e^2 e^{-(\varepsilon \cos x)}(\varepsilon \cos x) \\
&\qquad + e^{\varepsilon \cos x}(\varepsilon \cos x)\} \cos x
\end{aligned}$$

After integration we get

$$I_1 = -\left\{e^2 e^{-y} + e^y - e^2 e^{-y}(y+1) + e^y(y-1)\right\}\Big|_{\varepsilon\cos x}^{1+\zeta(x)} \cdot \sin x$$

or

$$I_1 = -\left\{e \cdot 2\zeta(1+\zeta) + [e^2 - 1 - (e^2+1)\,\varepsilon\cos x]\,\varepsilon\cos x\right\} \cdot \sin x$$

Keeping the terms of $O(\varepsilon)$ and $O(\zeta)$ only we have

$$I_1 = -\left\{2e\zeta + (e^2-1)\,\varepsilon\cos x\right\} \cdot \sin x$$

Within this accuracy we get

$$I_2 = \frac{d\zeta}{dx} \cdot e \cdot (3 + 2\zeta)\cos x$$

and

$$I_3 = -\varepsilon \sin x \cdot (1 - e^2)\cos x$$

If we leave out the term I_1, we have equality $I_2 + I_3 = 0$, which, after integration gives $3\zeta + \zeta^2 = \varepsilon \cdot (e - 1/e)\cos x$, and after linearization

$$\zeta = \frac{\varepsilon}{3} \cdot \left(e - \frac{1}{e}\right)\cos x$$

where $e \approx 2.71828$. In this case $O(\zeta) \simeq O(\varepsilon)$.

Appendix 5: Young-Laplace's Equation

Young-Laplace's equation is a statement of normal stress balance for static fluids meeting at an interface

$$p_1 - p_2 = \sigma\left(\frac{1}{R_1} + \frac{1}{R_2}\right)$$

where $p_1 - p_2$ is the pressure difference across the fluid interface, σ is the surface tension, and R_1 and R_2 are the principal radii of the interface surface curvature. In our case $p_1 = p$ is the pressure inside of the liquid, and $p_2 = p_A$ is the atmospheric pressure. Young-Laplace's equation can be written as

$$\frac{p_1}{\breve{p}} - \frac{p_2}{\breve{p}} = \frac{\sigma}{\breve{p}h}\left(\frac{h}{R_1} + \frac{h}{R_2}\right)$$

The value of surface tension of water against air $\sigma = 72.5\,\text{g/s}^2$, and in our natural units, cf. the Appendix 1,

$$\widetilde{\sigma} = \frac{\sigma}{\breve{p}h} = \frac{72.5}{10^5 \cdot \sin\alpha} \approx 7.25$$

For the droplet of radius 1 mm, $p - p_\text{A} = 0.0014\,\text{atm}$, and for the droplet of radius 0.1 mm, $p - p_\text{A} = 0.0144\,\text{atm}$, with p meaning the pressure inside of the liquid, and p_A is the atmospheric pressure.

References

1. G.A. Chechkin, A. Friedman, A.L. Piatnitski, The boundary-value problem in domains with very rapidly oscillating boundary. J. Math. Anal. Appl. **231**(1), 213–234 (1999)
2. R. Czapla, V.V. Mityushev, W. Nawalaniec, *Macroscopic Conductivity of Curvilinear Channels* (Department of Computer Science and Computational Methods, Pedagogical University of Cracow, Kraków, 2017, preprint)
3. F.M. Esquivelzeta-Rabell, B. Figueroa-Espinoza, D. Legendre, P. Salles, A note on the onset of recirculation in a 2D Couette flow over a wavy bottom. Phys. Fluids **27**(1), 014108-1–014108-14 (2015)
4. K. Evangelos, The impact of vegetation on the characteristics of the flow in an inclined open channel using the piv method. Water Resour. Ocean Sci. **1**(1), 1–6 (2012)
5. W. Huai, W. Wang, Y. Zeng, Two-layer model for open channel flow with submerged flexible vegetation. J. Hydraul. Res. **51**(6), 708–718 (2013)
6. E. Kubrak, J. Kubrak, P.M. Rowiński, Application of one-dimensional model to calculate water velocity distributions over elastic elements simulating Canadian waterweed plants (Elodea canadensis). Acta Geophys. **61**(1), 194–210 (2013)
7. L.D. Landau, E.M. Lifshitz, *Fluid Mechanics*. Course of Theoretical Physics (transl. from the Russian by J.B. Sykes, W.H. Reid), vol. 6, 2nd edn. (Pergamon Press, Oxford, 1987)
8. A.E. Malevich, V.V. Mityushev, P.M. Adler, Stokes flow through a channel with wavy walls. Acta Mech. **182**(3–4), 151–182 (2006)
9. A.E. Malevich, V.V. Mityushev, P.M. Adler, Couette flow in channels with wavy walls. Acta Mech. **197**(3), 247–283 (2008)
10. A.E. Malevich, V.V. Mityushev, P.M. Adler, Electrokinetic phenomena in wavy channels. J. Colloid Interface Sci. **345**, 72–87 (2010)
11. G. Sobin. *Luminous Debris: Reflecting on Vestige in Provence and Languedoc* (University of California Press, Berkeley, 2000)
12. R. Wojnar, W. Bielski, Flow in the canal with plants on the bottom. In *Complex Analysis and Potential Theory with Applications. Proceedings of the 9th ISAAC Congress, August, 2013, Kraków*, ed. by T.A. Azerogly, A. Golberg, S.V. Rogosin (Cambridge Scientific Publishers, Cambridge, 2014), pp. 167–183

Positive Solutions for a Nonlocal Resonant Problem of First Order

Mirosława Zima

Abstract We study a first order differential system subject to a nonlocal condition. Our goal in this paper is to establish conditions sufficient for the existence of positive solutions when the considered problem is at resonance. The key tool in our approach is Leggett-Williams norm-type theorem for coincidences due to O'Regan and Zima. We conclude the paper with several examples illustrating the main result.

Keywords Positive solution • Cone • Resonant problem • Nonlocal condition

Mathematics Subject Classification (2010) Primary 34B18; Secondary 34B10

1 Introduction

In the paper we study the existence of positive solutions for the following nonlocal problem of first order

$$\begin{cases} x'(t) = f(t, x(t)), \ t \in [0, T], \\ x(0) = \hat{\alpha}[x], \end{cases} \tag{1.1}$$

where $f : [0, T] \times \mathbb{R}^n_+ \to \mathbb{R}^n$ is a continuous function and $\hat{\alpha} : C([0, T]; \mathbb{R}^n) \to \mathbb{R}^n$ is a bounded linear functional. Nonlocal problems for first order differential equations, differential inclusions and systems of differential equations have been studied recently in [1–4, 7, 11, 16, 20, 21], and [24]. Observe that for $\hat{\alpha}[x] = x(T)$ we obtain a periodic boundary condition $x(0) = x(T)$. The problems of this kind were considered for example in [9, 15, 17, 22, 25, 26], and [27]. The methods applied for nonlocal problems include the monotone iterative method [11], the

M. Zima (✉)
Faculty of Mathematics and Natural Sciences, Department of Functional Analysis, University of Rzeszów, Pigonia 1, Rzeszów 35-959, Poland
e-mail: mzima@ur.edu.pl

P. Drygaś, S. Rogosin (eds.), *Modern Problems in Applied Analysis*,
Trends in Mathematics, https://doi.org/10.1007/978-3-319-72640-3_14

coincidence degree theory due to Mawhin [16], the Leggett-Williams multiple fixed point theorem and Krasnoselskii-Guo fixed point theorem of cone expansion and compression [17].

The purpose of this paper is to establish a new existence result for the nonlocal problem (1.1) at resonance, that includes the case of periodic boundary condition. For $x \in C([0,T];\mathbb{R}^n)$, $x = (x_1, x_2, \ldots, x_n)$ we will deal with $\hat{\alpha}$ of the form

$$\hat{\alpha}[x_1, x_2, \ldots, x_n] = (\alpha[x_1], \alpha[x_2], \ldots, \alpha[x_n]). \tag{1.2}$$

Here α is a positive continuous linear functional given by the Riemann-Stieltjes integral, that is, for a continuous function $\varphi : [0,T] \to \mathbb{R}$ we have

$$\alpha[\varphi] = \int_0^T \varphi(t)dA(t), \tag{1.3}$$

where we assume A is a function of bounded variation and dA is a positive measure. Observe that (1.3) includes multipoint and integral boundary conditions as the special cases.

Our approach is to write (1.1) as the coincidence equation $Lx = Nx$, where L is a Fredholm operator of index zero and N is a nonlinear operator, and to apply Leggett-Williams norm-type theorem for coincidences due to O'Regan and Zima [22]. Usually, when the coincidence technique is employed to study the resonant problems, L is chosen as a linear differential operator, and N is the Nemytskii operator, see for example [10, 12–14, 18, 22], and [28]. Here, similarly to [6], we put $Lx(t) = x(t) - \hat{\alpha}[x]$, and N is a nonlinear integral operator of Volterra type. To the best of our knowledge, this approach is used for (1.1) for the first time. Our method enables us to complement some results obtained for first order systems and scalar resonant problems in the recent literature.

2 Positive Solutions

We begin this section by providing some background on cone theory and Fredholm operators in Banach spaces. We recall that a cone C in a real Banach space X is a nonempty closed convex set such that $C \neq \{0\}$, $\lambda x \in C$ for all $x \in C$ and $\lambda \geq 0$, and $C \cap (-C) = \{0\}$.

Moreover, the following property holds for every cone in a Banach space.

Lemma 2.1 ([23]) *For every $u \in C \setminus \{0\}$ there exists a positive number $\sigma(u)$ such that*

$$\|x + u\| \geq \sigma(u)\|x\|$$

for all $x \in C$.

Let X and Y be Banach spaces. Consider a linear mapping $L : \operatorname{dom} L \subset X \to Y$ and a nonlinear operator $N : X \to Y$. We say that L is a Fredholm operator of index zero if $\operatorname{Im} L$ is closed and $\dim \operatorname{Ker} L = \operatorname{codim} \operatorname{Im} L < \infty$. If L is Fredholm of index zero, then there exist continuous projections $P : X \to X$ and $Q : Y \to Y$ such that $\operatorname{Im} P = \operatorname{Ker} L$ and $\operatorname{Ker} Q = \operatorname{Im} L$ (see for example [19, 25]). Moreover, since $\dim \operatorname{Im} Q = \operatorname{codim} \operatorname{Im} L$, there exists an isomorphism $J : \operatorname{Im} Q \to \operatorname{Ker} L$. Denote by L_P the restriction of L to $\operatorname{Ker} P \cap \operatorname{dom} L$. Clearly, L_P is an isomorphism from $\operatorname{Ker} P \cap \operatorname{dom} L$ to $\operatorname{Im} L$ and its inverse

$$K_P : \operatorname{Im} L \to \operatorname{Ker} P \cap \operatorname{dom} L$$

is defined. Then the equation $Lx = Nx$ is equivalent to $x = \Psi x$, where

$$\Psi = P + JQN + K_P(I - Q)N.$$

Let $\rho : X \to C$ be a retraction, that is, a continuous mapping such that $\rho(x) = x$ for all $x \in C$. Put

$$\Psi_\rho = \Psi \circ \rho.$$

The following theorem due to O'Regan and Zima will be applied in order to establish the existence result for (1.1).

Theorem 2.2 ([22]) *Let Ω_1, Ω_2 be open bounded subsets of X with $\overline{\Omega}_1 \subset \Omega_2$ and $C \cap (\overline{\Omega}_2 \setminus \Omega_1) \neq \emptyset$. Assume that:*

1° *L is a Fredholm operator of index zero,*
2° *$QN : X \to Y$ is continuous and bounded and $K_P(I - Q)N : X \to X$ is compact on every bounded subset of X,*
3° *$Lx \neq \lambda Nx$ for all $x \in C \cap \partial\Omega_2 \cap \operatorname{dom} L$ and $\lambda \in (0, 1)$,*
4° *ρ maps subsets of $\overline{\Omega}_2$ into bounded subsets of C,*
5° *$d_B([I - (P + JQN)\rho]|_{\operatorname{Ker} L}, \operatorname{Ker} L \cap \Omega_2, 0) \neq 0$, where d_B stands for the Brouwer degree,*
6° *there exists $u_0 \in C \setminus \{0\}$ such that $\|x\| \leq \sigma(u_0)\|\Psi x\|$ for $x \in C(u_0) \cap \partial\Omega_1$, where*

$$C(u_0) = \{x \in C : x - \eta u_0 \in C \text{ for some } \eta > 0\}$$

and $\sigma(u_0)$ is such that $\|x + u_0\| \geq \sigma(u_0)\|x\|$ for every $x \in C$,
7° *$(P + JQN)\rho(\partial\Omega_2) \subset C$ and $\Psi_\rho(\overline{\Omega}_2 \setminus \Omega_1) \subset C$.*

Then the equation $Lx = Nx$ has a solution in the set $C \cap (\overline{\Omega}_2 \setminus \Omega_1)$.

It is worth mentioning that Theorem 2.2 is a modification of Mawhin's coincidence degree theory [19]. For similar results see for example [5, 8] and [25].

We now give some notations which are used throughout the paper. We set:

$$k(t,s) = \begin{cases} 1, & 0 \le s \le t \le T, \\ 0, & 0 \le t < s \le T, \end{cases} \tag{2.1}$$

$$K(s) = \int_0^T k(t,s)dA(t), s \in [0,T], \tag{2.2}$$

$$G(t,s) = \begin{cases} \frac{s}{T}, & 0 \le s \le t \le T, \\ \frac{s}{T} - 1, & 0 \le t < s \le T, \end{cases} \tag{2.3}$$

and

$$\widetilde{G}(t,s) = MK(s) + G(t,s), \tag{2.4}$$

where M is a positive constant. For the vectors $u, v \in \mathbb{R}^n$, $u = (u_1, u_2, \dots, u_n)$, $v = (v_1, v_2, \dots, v_n)$, $\langle u, v \rangle$ denotes the standard inner product in $\mathbb{R}^n$ and $\|u\|$ denotes the Euclidean norm of u in $\mathbb{R}^n$. Moreover, we let $u \le v$ (and $u < v$) if $u_i \le v_i$ ($u_i < v_i$, respectively) for $i = 1, 2, \dots, n$.

Throughout the paper we assume:

(H1) $f : [0,T] \times \mathbb{R}^n_+ \to \mathbb{R}^n$ is continuous,
(H2) dA is a positive measure and $\alpha[1] = 1$.

Clearly, $\alpha[1] = 1$ reads as $\int_0^T dA(s) = 1$ and it is equivalent to $\hat{\alpha}[\hat{1}] = \hat{1}$, where $\hat{1}$ denotes the constant vector function $(1, 1, \dots, 1)$. It corresponds to the resonant case, that is, the homogeneous problem

$$\begin{cases} x'(t) = 0, \ t \in [0,T], \\ x(0) = \hat{\alpha}[x], \end{cases}$$

associated with (1.1) has nontrivial solutions. It is important to notice that (H2), (2.1) and (2.2) imply $K(s) \ge 0$ on $[0,T]$.

We also assume that there exist $R > 0$, $r \in (0,R)$, $M > 0$, $\mu \in (0,1)$ and continuous functions $g_i : [0,T] \to [0,\infty)$ and $h_i : [0,r] \to [0,\infty)$, $i = 1, 2, \dots, n$ such that:

(H3) K is not identically zero on any subinterval of $[0,T]$ and $1 - \mu M \int_0^T K(\tau)d\tau \ge 0$,
(H4) $\frac{1}{T} - \mu MK(s) \ge 0$ for $s \in [0,T]$,
(H5) $\widetilde{G}(t,s) \ge 0$ and $\frac{1}{T} - \mu \widetilde{G}(t,s) \ge 0$ for $t, s \in [0,T]$,
(H6) $\langle x, f(t,x) \rangle < 0$ for $t \in [0,T]$ and $x \ge 0$ with $\|x\| = R$,
(H7) $f(t,x) > -\mu x$ for $t \in [0,T]$ and $x \ge 0$ with $\|x\| \in [0,R]$,

(H8) $f_i(t, x_1, x_2, \ldots, x_n) \geq g_i(t)h_i(x_i)$ for $t \in [0, T]$, $x_1, x_2, \ldots, x_n \in [0, r]$, $h_i(x_i)/x_i$ is decreasing on $(0, r]$.

Theorem 2.3 *Under the assumptions (H1)–(H8), problem* (1.1) *has a positive solution on* $[0, T]$.

Proof Consider the Banach spaces $X = Y = C([0, T]; \mathbb{R}^n)$ endowed with the standard supremum norm

$$\|x\|_\infty = \sup_{t \in [0,T]} \|x(t)\|,$$

where

$$\|x(t)\| = \sqrt{\sum_{i=1}^{n} x_i^2(t)}.$$

Let

$$Lx(t) = x(t) - \hat{\alpha}[x], \quad t \in [0, T],$$

for $x \in \operatorname{dom} L = X$, and

$$Nx(t) = \int_0^t f(\tau, x(\tau))d\tau, \quad t \in [0, T],$$

for $x \in X$. It is easy to see that every solution of the coincidence equation

$$Lx = Nx,$$

is a solution of (1.1). In view of (H2) we get

$$\operatorname{Ker} L = \{x \in \operatorname{dom} L : x(t) = c,\ t \in [0, T],\ c \in \mathbb{R}^n\}$$

and

$$\operatorname{Im} L = \{y \in Y : \hat{\alpha}[y] = 0\}.$$

Clearly, $\operatorname{Im} L$ is closed and $Y = \widetilde{Y} + \operatorname{Im} L$ with

$$\widetilde{Y} = \{\widetilde{y} \in \widetilde{Y} : \widetilde{y}(t) = \hat{\alpha}[y],\ y \in Y\}.$$

Observe that $\widetilde{Y} \cap \operatorname{Im} L = \{0\}$, and therefore $Y = \widetilde{Y} \oplus \operatorname{Im} L$. Moreover, since $\dim \widetilde{Y} = n$, we have $\operatorname{codim} \operatorname{Im} L = n$. Thus L is Fredholm of index zero, and the assumption 1° is satisfied.

Next, define the projections $P : X \to X$ and $Q : Y \to Y$ by

$$Px = \frac{1}{T}\int_0^T x(s)ds, \quad t \in [0, T],$$

and

$$Qy = \hat{\alpha}[y],$$

respectively. Standard calculations show that for $y \in \operatorname{Im} L$ the inverse K_P of L_P is given by

$$K_P y(t) = y(t) - \frac{1}{T}\int_0^T y(\tau)d\tau, \quad t \in [0, T].$$

It follows from (H1) that 2° is fulfilled.

For $z \in \operatorname{Im} Q$ define

$$Jz(t) = Mz(t).$$

Then J is an isomorphism from $\operatorname{Im} Q$ to $\operatorname{Ker} L$. Next, consider a cone

$$C = \{x \in X : x(t) \geq 0 \text{ on } [0, T]\}.$$

For $u_0 = \hat{1}$ we have $\sigma(u_0) = 1$ and

$$C(u_0) = \{x \in C : x(t) > 0 \text{ on } [0, T]\}.$$

Let

$$m = \left(\sum_{i=1}^{n}\left(1 + \frac{h_i(r)}{r}\int_0^T \widetilde{G}(T, s)g_i(s)ds\right)^2\right)^{-\frac{1}{2}}, \tag{2.5}$$

$$\Omega_1 = \{x \in X : \|x\|_\infty < r,\ x_i(t) > m\|x\|_\infty \ \text{ on } \ [0, T], i = 1, 2, \ldots, n\},$$

and

$$\Omega_2 = \{x \in X : \|x\|_\infty < R\}.$$

Obviously, $m \in (0, 1)$, and Ω_1 and Ω_2 are open bounded subsets of X, and $\overline{\Omega}_1 \subset \Omega_2$.

To verify 3° suppose that there exist $x_0 \in C \cap \partial\Omega_2 \cap \operatorname{dom} L$ and $\lambda_0 \in (0, 1)$ such that $Lx_0 = \lambda_0 Nx_0$. Then, in particular $x_0(t) \geq 0$ on $[0, T]$, $\|x_0\|_\infty = R$, and

$$x_0'(t) = \lambda_0 f(t, x_0(t)), \quad t \in [0, T]. \tag{2.6}$$

This implies

$$\frac{1}{2}(\|x_0(t)\|^2)' = \langle x_0(t), x_0'(t)\rangle = \lambda_0 \langle x_0(t), f(t, x_0(t))\rangle.$$

We consider two cases. If $\|x_0\|_\infty = \|x_0(t_0)\| = R$ for some $t_0 \in (0, T]$, then

$$0 \le \frac{1}{2}(\|x_0(t)\|^2)'|_{t=t_0} = \lambda_0 \langle x_0(t_0), f(t_0, x_0(t_0))\rangle,$$

in contradiction with (H6). If $\|x_0\|_\infty = \|x_0(t_0)\| = R$ only for $t_0 = 0$, then (H2) gives

$$\begin{aligned} R^2 &= \|x_0(0)\|^2 = \|\hat{\alpha}[x_0]\|^2 = \sum_{i=1}^{n} \alpha^2[x_{0i}] = \sum_{i=1}^{n} \Big(\int_0^T x_{0i}(t)dA(t)\Big)^2 \\ &\le \sum_{i=1}^{n} \Big(\int_0^T dA(t) \int_0^T x_{0i}^2(t)dA(t)\Big) = \sum_{i=1}^{n} \int_0^T x_{0i}^2(t)dA(t) \\ &= \int_0^T (\sum_{i=1}^{n} x_{0i}^2(t))dA(t) < \int_0^T (\sum_{i=1}^{n} x_{0i}^2(0))dA(t) = R^2, \end{aligned}$$

a contradiction. Next, for $x \in X$ define

$$\rho x(t) = |x(t)|.$$

Evidently, ρ is a retraction and maps subsets of $\overline{\Omega}_2$ into bounded subsets of C, so 4° holds. To verify 5° we consider the mapping

$$H(x, \lambda)(t) = x(t) - \lambda\Big(\frac{1}{T}\int_0^T |x(s)|ds + M\int_0^T \int_0^t f(\tau, |x(\tau)|)d\tau\, dA(t)\Big)$$

for $x \in \operatorname{Ker} L \cap \Omega_2$ and $\lambda \in [0, 1]$. Then

$$\begin{aligned} H(x, \lambda)(t) &= x(t) - \lambda\Big(\frac{1}{T}\int_0^T |x(s)|ds + M\int_0^T \int_0^T k(t, \tau)f(\tau, |x(\tau)|)d\tau\, dA(t)\Big) \\ &= x(t) - \lambda\Big(\frac{1}{T}\int_0^T |x(s)|ds + M\int_0^T K(\tau)f(\tau, |x(\tau)|)d\tau\Big). \end{aligned}$$

Observe that if $x \in \operatorname{Ker} L \cap \Omega_2$, then $x(t) = c \in \mathbb{R}^2$ on $[0, T]$ and $\|x\|_\infty < R$. Suppose $H(x, \lambda) = 0$ for $x \in \partial\Omega_2$. Then $\|c\| = R$ and (H3) and (H7) imply

$$\begin{aligned} c &= \lambda\Big(\frac{1}{T}\int_0^T |c|ds + M\int_0^T K(s)f(s, |c|)ds\Big) \\ &\ge \lambda\Big(1 - \mu M\int_0^T K(s)ds\Big)|c| \ge 0. \end{aligned}$$

Therefore $|c| = c$, and we get

$$R^2 = \langle c, c\rangle = \lambda\Big(\frac{1}{T}\int_0^T \langle c, c\rangle ds + M\int_0^T K(s)\langle c, f(s, c)\rangle ds\Big).$$

This in view of (H3) and (H6) leads to the contradiction

$$0 \le R^2(1-\lambda) = \lambda M\int_0^T K(s)\langle c, f(s, c)\rangle ds < 0.$$

Thus $H(x, \lambda) \neq 0$ for $x \in \partial\Omega_2$ and $\lambda \in [0, 1]$. This implies

$$d_B(H(x, 0), \operatorname{Ker} L \cap \Omega_2, 0) = d_B(H(x, 1), \operatorname{Ker} L \cap \Omega_2, 0),$$

and

$$d_B\big([I - (P + JQN)\rho]\big|_{\operatorname{Ker} L}, \operatorname{Ker} L \cap \Omega_2, 0\big) = d_B(H(c, 1), \operatorname{Ker} L \cap \Omega_2, 0) \neq 0.$$

We next show that 6° holds. Let $x \in C(u_0) \cap \partial\Omega_1$. Then for $t \in [0, T]$ we have in particular $r \ge x_i(t) \ge m\|x\|_\infty > 0$, and by (H5) and (H8) we obtain for $i = 1, 2, \ldots, n$

$$\begin{aligned}
(\Psi x)_i(T) &= \frac{1}{T}\int_0^T x_i(s)ds + M\int_0^T K(s)f_i(s, x_1(s), x_2(s), \ldots, x_n(s))ds \\
&\quad + \int_0^T G(T, s)f_i(s, x_1(s), x_2(s), \ldots, x_n(s))ds \\
&= \frac{1}{T}\int_0^T x_i(s)ds + \int_0^T (MK(s) + G(T, s))f_i(s, x_1(s), x_2(s), \ldots, x_n(s))ds \\
&= \frac{1}{T}\int_0^T x_i(s)ds + \int_0^T \widetilde{G}(T, s)f_i(s, x_1(s), x_2(s), \ldots, x_n(s))ds \\
&\ge \frac{1}{T}\int_0^T x_i(s)ds + \int_0^T \widetilde{G}(T, s)g_i(s)\frac{h_i(x_i(s))}{x_i(s)}x_i(s)ds \\
&\ge \frac{1}{T}\int_0^T x_i(s)ds + \frac{h_i(r)}{r}\int_0^T \widetilde{G}(T, s)g_i(s)x_i(s)ds \\
&\ge m\|x\|_\infty + m\|x\|_\infty\frac{h_i(r)}{r}\int_0^T \widetilde{G}(T, s)g_i(s)ds \\
&= m\left(1 + \frac{h_i(r)}{r}\int_0^T \widetilde{G}(T, s)g_i(s)ds\right)\|x\|_\infty.
\end{aligned}$$

This implies

$$\sup_{t\in[0,T]} \|(\Psi x)(t)\| \geq \|(\Psi x)(T)\| \geq m\sqrt{\sum_{i=1}^{n}\left(1+\frac{h_i(r)}{r}\int_0^T \widetilde{G}(T,s)g_i(s)ds\right)^2}\|x\|_\infty \geq \|x\|_\infty,$$

and therefore $\|x\|_\infty \leq \|\Psi x\|_\infty$ for $x \in C(u_0) \cap \partial\Omega_1$, so 6° is satisfied.

If $x \in \partial\Omega_2$ then in view of (H3), (H4) and (H7) we get

$$\begin{aligned}(P+JQN)(\rho x)(t) &= \frac{1}{T}\int_0^T |x(s)|ds + M\int_0^T K(s)f(s,|x(s)|)ds\\ &\geq \frac{1}{T}\int_0^T |x(s)|ds - \mu M\int_0^T K(s)|x(s)|ds\\ &= \int_0^T \left(\frac{1}{T} - \mu MK(s)\right)|x(s)|ds \geq 0.\end{aligned}$$

Moreover, for $x \in \overline{\Omega}_2 \setminus \Omega_1$ we have from (H5) and (H7)

$$\begin{aligned}\Psi_\rho x(t) &= \frac{1}{T}\int_0^T |x(s)|ds + M\int_0^T K(s)f(s,|x(s)|)ds + \int_0^T G(t,s)f(s,|x(s)|)ds\\ &= \frac{1}{T}\int_0^T |x(s)|ds + \int_0^T \widetilde{G}(t,s)f(s,|x(s)|)ds\\ &\geq \frac{1}{T}\int_0^T |x(s)|ds - \mu\int_0^T \widetilde{G}(t,s)|x(s)|ds\\ &= \int_0^T \left(\frac{1}{T} - \mu\widetilde{G}(t,s)\right)|x(s)|ds \geq 0.\end{aligned}$$

Thus, 7° is fulfilled and the assertion follows. □

Remark 2.4 In [20] the author dealt with the following nonlocal problem

$$\begin{cases} x_1'(t) = f_1(t,x_1,x_2),\\ x_2'(t) = f_2(t,x_1,x_2),\\ x_1(0) = \alpha_1[x_1],\\ x_2(0) = \alpha_2[x_2]. \end{cases} \tag{2.7}$$

The Perov, Schauder and Leray-Schauder theorems were used to study the uniqueness and existence for (2.7). A key assumption in [20] is $\alpha_1[1] \neq 1$ and $\alpha_2[1] \neq 1$. Therefore, in the case $\alpha = \alpha_1 = \alpha_2$, and $\alpha[1] = 1$ as assumed in (H2), Theorem 2.3 complements the results from [20].

3 Some Examples

In this section we provide a few examples illustrating the assumptions that appear in Theorem 2.3. Some computations were made with the software system Maple. We begin with an example involving an integral boundary condition.

Example Consider the following scalar problem

$$\begin{cases} x'(t) = (t^2 + 0.25)(x^2 - 3x + 0.5 + 0.1\sqrt{t+x}), \ t \in [0, 1], \\ x(0) = \int_0^1 x(s)ds. \end{cases} \tag{3.1}$$

In this case we have $f(t, x) = (t^2 + 0.25)(x^2 - 3x + 0.5 + 0.1\sqrt{t+x})$ and $\alpha[x] = \int_0^1 x(s)ds$. By (2.2) and (2.3) we get $K(s) = 1 - s$ for $s \in [0, 1]$, and

$$G(t, s) = \begin{cases} s, & 0 \le s \le t \le 1, \\ s - 1, & 0 \le t < s \le 1. \end{cases}$$

It is easy to show that the assumptions (H3)–(H7) of Theorem 2.3 are satisfied with $M = 1$, $\mu = 0.95$ and $R = 0.3$. In particular, since $f(t, 0.3) < 0$ for $t \in [0, 1]$, (H6) holds. Moreover, for $r = 0.15$, $g(t) = t^2 + 0.25$, and $h(x) = x^2 - 3x + 0.5$, (H8) is fulfilled. By Theorem 2.3, problem (3.1) has a solution, positive on $[0, 1]$.

Next example deals with the system of two equations subject to periodic boundary condition.

Example We will show that the following periodic problem

$$\begin{cases} x_1'(t) = (t + 1)((x_1 + x_2)^2 - 3(x_1 + x_2) + 1 + \sin x_2), \\ x_2'(t) = (t + 1)((x_1 + x_2)^2 - 2(x_1 + x_2) + 0.3 + \sin x_1), \\ x_1(0) = x_1(1), \\ x_2(0) = x_2(1), \end{cases} \tag{3.2}$$

has a positive solution on $[0, 1]$. We have $\alpha[x_1] = x_1(1)$, $\alpha[x_2] = x_2(1)$, and $K(s) = 1$. The assumptions of Theorem 2.3 are fulfilled with $R = 0.4$, $r = 0.1$, $M = 1$, $\mu = 0.95$, $g_1(t) = g_2(t) = t + 1$, $h_1(t) = 0.7(x_1^2 - 3x_1 + 1)$, and $h_2(t) = 0.2(x_2^2 - 2x_2 + 0.3)$. In particular, using the Maple software, we can show that for $x_1, x_2 \ge 0$ and $\sqrt{x_1^2 + x_2^2} = 0.4$

$$x_1((x_1+x_2)^2-3(x_1+x_2)+1+\sin x_2)+x_2((x_1+x_2)^2-2(x_1+x_2)+0.3+\sin x_1) < 0.$$

Hence for $t \in [0, 1]$, $x_1, x_2 \geq 0$ and $\sqrt{x_1^2 + x_2^2} = 0.4$ we get

$$\langle x, f(t,x)\rangle = (t+1)\big(x_1((x_1+x_2)^2 - 3(x_1+x_2) + 1 + \sin x_2) \\ + x_2((x_1+x_2)^2 - 2(x_1+x_2) + 0.3 + \sin x_1)\big) < 0,$$

so (H6) is fulfilled. The application of Theorem 2.3 yields the result.

We finally discuss an example with a multipoint boundary condition.

Example Let us consider problem (1.1) with $\hat{\alpha}[x] = ax(\xi) + (1-a)x(T)$, where $\xi \in (0, T)$ and $a \in (0, 1)$. We will derive the conditions that imply (H3)–(H5). From (2.2) we obtain

$$K(s) = \begin{cases} 1, & 0 \leq s \leq \xi, \\ 1-a, & \xi < s \leq T. \end{cases}$$

Since $\int_0^T K(\tau)d\tau = \xi + (1-a)(T-\xi)$, in order to satisfy (H3) we need

$$M \leq \frac{1}{\mu(\xi + (1-a)(T-\xi))}.$$

On the other hand, (H4) holds if $M \leq \frac{1}{\mu T}$, while (H5) is fulfilled if $M \geq \frac{1}{1-a}$ and $M \leq \frac{1}{\mu T} - 1$. Thus (H3)–(H5) are satisfied if

$$\mu T < 1, \qquad \frac{1}{1-a} < \min\left\{\frac{1}{\mu(\xi + (1-a)(T-\xi))}, \frac{1}{\mu T} - 1\right\},$$

and

$$M \in \left[\frac{1}{1-a}, \min\left\{\frac{1}{\mu(\xi + (1-a)(T-\xi))}, \frac{1}{\mu T} - 1\right\}\right].$$

Acknowledgements This work was completed with the partial support of the Centre for Innovation and Transfer of Natural Science and Engineering Knowledge of University of Rzeszów. The author wishes to express her thanks to the referees for careful reading of the manuscript and constructive comments.

References

1. D.R. Anderson, Existence of three solutions for a first-order problem with nonlinear nonlocal boundary conditions. J. Math. Anal. Appl. **408**, 318–323 (2013)
2. O. Bolojan, R. Precup, Implicit first order differential systems with nonlocal conditions. Electron. J. Qual. Theor. Differ. Equ. **2014**(69), 1–13 (2014)

3. O. Bolojan-Nica, G. Infante, R. Precup, Existence results for systems with coupled nonlocal initial conditions. Nonlinear Anal. **94**, 231–242 (2014)
4. A. Boucherif, First-order differential inclusions with nonlocal initial conditions. Appl. Math. Lett. **15**, 409–414 (2002)
5. C.T. Cremins, A fixed point index and existence theorems for semilinear equations in cones. Nonlinear Anal. **46**, 789–806 (2001)
6. D. Franco, G. Infante, M. Zima, Second order nonlocal boundary value problems at resonance. Math. Nachr. **284**, 875–884 (2011)
7. D. Franco, J.J. Nieto, D. O'Regan, Existence of solutions for first order ordinary differential equations with nonlinear boundary conditions. Appl. Math. Comput. **153**, 793–802 (2004)
8. R.E. Gaines, J. Santanilla, A coincidence theorem in convex sets with applications to periodic solutions of ordinary differential equations. Rocky Mt. J. Math. **12**, 669–678 (1982)
9. J.R. Graef, S. Padhi, S. Pati, Periodic solutions of some models with strong Allee effects. Nonlinear Anal. Real World Appl. **13**, 569–581 (2012)
10. G. Infante, M. Zima, Positive solutions of multi-point boundary value probelms at resonance. Nonlinear Anal. **69**, 2458–2465 (2008)
11. T. Jankowski, Boundary value problems for first order differential equations of mixed type. Nonlinear Anal. **64**, 1984–1997 (2006)
12. W. Jiang, C. Yang, The existence of positive solutions for multi-point boundary value problem at resonance on the half-line. Bound. Value Probl. **2016**, 13 (2016)
13. N. Kosmatov, Multi-point boundary value problems on an unbounded domain at resonance. Nonlinear Anal. **68**, 2158–2171 (2008)
14. N. Kosmatov, A singular non-local problem at resonance. J. Math. Anal. Appl. **394**, 425–431 (2012)
15. V. Lakshmikantham, S. Leela, Existence and monotone method for periodic solutions of first order differential equations. J. Math. Anal. Appl. **91**, 237–243 (1983)
16. B. Liu, Existence and uniqueness of solutions to first-order multipoint boundary value problems. Appl. Math. Lett. **17**, 1307–1316 (2004)
17. Y. Liu, Multiple solutions of periodic boundary value problems for first order differential equations. Comput. Math. Appl. **54**, 1–8 (2007)
18. B. Liu, Z. Zhao, A note on multi-point boundary value problems. Nonlinear Anal. **67**, 2680–2689 (2007)
19. J. Mawhin, Equivalence theorems for nonlinear operator equations and coincidence degree theory for mappings in locally convex topological vector spaces. J. Differ. Equ. **12**, 610–636 (1972)
20. O. Nica, Nonlocal initial value problems for first order differential systems. Fixed Point Theory **13**, 603–612 (2012)
21. J.J. Nieto, R. Rodríguez-López, Greens function for first-order multipoint boundary value problems and applications to the existence of solutions with constant sign. J. Math. Anal. Appl. **388**, 952–963 (2012)
22. D. O'Regan, M. Zima, Leggett-Williams norm-type theorems for coincidences. Arch. Math. **87**, 233–244 (2006)
23. W.V. Petryshyn, On the solvability of $x \in Tx + \lambda Fx$ in quasinormal cones with T and F k-set contractive. Nonlinear Anal. **5**, 585–591 (1981)
24. R. Precup, D. Trif, Multiple positive solutions of non-local initial value problems for first order differential systems. Nonlinear Anal. **75**, 5961–5970 (2012)
25. J. Santanilla, Some coincidence theorems in wedges, cones, and convex sets. J. Math. Anal. Appl. **105**, 357–371 (1985)
26. J. Santanilla, Nonnegative solutions to boundary value problems for nonlinear first and second order ordinary differential equations. J. Math. Anal. Appl. **126**, 397–408 (1987)
27. C.C. Tisdell, Existence of solutions to first-order periodic boundary value problems. J. Math. Anal. Appl. **323**, 1325–1332 (2006)
28. M. Zima, P. Drygaś, Existence of positive solutions for a kind of periodic boundary value problem at resonance. Bound. Value Probl. **2013**, 19 (2013)

Printed in the United States
By Bookmasters